Horticultural Crops
and
Fruit Flies of India

Horticultural Crops and Fruit Flies of India

NEERJA AGRAWAL
Emeritus Professor
Department of Entomology
CSA University of Agriculture and Technology
Kanpur-208002, **Uttar Pradesh, INDIA**

KALYANI PUBLISHERS
LUDHIANA • NEW DELHI • NOIDA (U.P.) • HYDERABAD • CHENNAI
KOLKATA • CUTTACK • GUWAHATI • KOCHI • BENGALURU

KALYANI PUBLISHERS

Head Office

B-1/1292, Rajinder Nagar, **Ludhiana-**141 008 • Ph : 0161-2760031, 2745756
E-mail : kalyanibooks@yahoo.co.in **Website :** www.kalyanipublishers.co.in

Administration Office

4779/23, Ansari Road, Daryaganj, New Delhi-110 002 • Ph : 011-23271469, 23274393
E-mail : kalyani_delhi@yahoo.co.in, kalyani_delhi@kalyanipublishers.in

Works

B-16, Sector-8, **NOIDA** (U.P.)

Branches

1, Mahalakshmi Street, T. Nagar, **Chennai-**600 017 • Ph : 044-24344684
Gopabandhu Lane, Behind Govt. Bus Stand, Badambadi,
Cuttack-753 012 (Odisha) • Ph : 0671-2311391
3-5-1108, Narayanaguda, **Hyderabad-**500 029 • Ph : 040-24750368
10/2B, Ramanath Mazumdar Street, **Kolkata-**700 009 • Ph : 033-22416024
Arunalaya, 1st Floor, Saraswati Road, Pan Bazar, **Guwahati-**781 001 • Ph : 0361-2731274
Koratti Parambil House, Convent Road, **Kochi-**682 035 • Ph : 0484-2367189
No. 24 & 25, 1st Floor, Hameed Shah Complex, Cubbonpet Main Road, **Bengaluru-**560 002

10 + 174 = 184 Pages 28" x 40" 11.5 Formes

KPP P 30992 05

Laser Typesetting at
Elite-Art

ISBN 978-93-89900-62-0

PRINTED IN INDIA
at Diamond Agencies Pvt. Ltd., B-125, Sec. 63, NOIDA

चन्द्रशेखर आजाद कृषि एवं प्रौद्योगिक विश्वविद्यालय
कानपुर-208 002, उत्तर प्रदेश, भारत
Chandra Shekhar Azad University of Agriculture & Technology
Kanpur-208002, Uttar Pradesh, India

डा० एस० सोलोमन
कुलपति
Dr. S. Solomon
Vice-Chancellor

Dated : November 01, 2019

FOREWORD

It gives me immense pleasure to write foreword of the book **"Horticultural Crops and Fruit Flies of India."** It is an effort by the author to provide a well written document for the production of quality fruits and vegetables with measures for integrated approach to manage fruit flies and enhance export.

India being one the major producers of fruits and vegetables in the world, the responsibility to protect our horticultural produce increases many folds with increasing globalization. Man considers insects as his enemy and to safeguard his interest tries to kill them by various means. But without the knowledge of the especially fruit flies the struggle goes on a never ending ordeal.

The book is written in all the minute details about fruit flies which are noxious pests in India; and most precisely, are of quarantine importance.

I am sure; the book will provide a good knowledge to farmers, fruit and vegetable growers; researchers; teachers, entomologists, plant protection and extension workers. I believe that the book will get a good response from students and teachers connected with agriculture and its related discipline especially Entomology.

(Sushil Solomon)
Vice-Chancellor

Preface

In view of achieving healthy food for the increasing population of men, area under cultivation has got momentum for maximum food production. In doing so, fruits and vegetables gained importance as being rich in vitamins and minerals, which are essential for a healthy dietary requirement. The insects evolved much earlier than man, around 250 million years ago while the existence of man dates back to 1.0 million years. In order to meet his requirement of food and shelter, the natural habitat of insects have been destroyed, making them pests of various field crops and orchards.

Throughout the past century, an increasing amount of scientific research work has been carried out to protect crops from diseases and insects; and horticultural crops are no exception. Man is aware that by conquering the pests, the healthy environment of earth is being depleted. Fruit fly (Tephritidae) is such noxious pest that it has gained the attention as a Key pest. The scope of fruit fly and their management enlarged tremendously during last few years. An important study on biology and its management in this discipline has gained its orientation to global problems which draw attention of scientists, researchers and plant protection specialists to solve the problem of fruit fly on horticultural crops.

The changing scenario in fruit fly insect pest complex of agricultural and horticultural crops, introduction of a number of new classes of pesticides, Para-pheromones and management tactics developed for controlling the pest species for eco-friendly management have made positive impact on crop protection. In the field of fruit fly, as a part and their management, significant progress has been made.

Since, fruit flies are economically important pest of many fruits and vegetables in India; they are serious impediment to fruits and vegetable cultivation in our country. Due to inadequate knowledge of how to use this information for trade negotiations, quarantine or for management purpose, the book "Horticultural Crops and Fruit flies of India" has been written by compiling all available sources and literature on bio -diversity, biology, prevalence and abundance of fruit flies and their management. Thus, in the present book efforts have been made to review the work done on relationship between fruit flies and weather parameters, seasonal activity and population dynamics, responses to Para pheromones, attractants, food baits, and colour of traps and species composition of this noxious pest; as studied during an intensive research. The life history, habits of fruit fly and possible eco-friendly measures to combat fruit fly menace, have also been incorporated in this book.

The idea behind writing this book is that there have been considerable changes in the scientific names of many economically important species and also addition of new pest species. The immense advancement in theoretical and applied aspects that have taken place in the past decade implied me to bring out a completely enlarged version of the present book.

I have striven hard to bring out this comprehensive and modern first edition of this compilation on fruit flies and its management with a view to meet the long felt need of the students, teachers, research workers, farmers and extension personnel in the country.

Since, I had been Area Coordinator of National Integrated Fruit fly Surveillance Project, I got an opportunity to adroit myself with fruit flies all over India, and this inspired me to document all possible literature and knowledge gained during the Research Project. I have been working on fruit flies since long and tried to bring this knowledge in black and white; and now it is high time to get it published. This book is the outcome of sincere, dedicated, long field research, training and extension experience of my almost 35 years of service. It is extensively written in simple and lucid language and has been updated with the useful illustrations and photographs.

I wish to mention that without support and constant encouragement of my daughters Charu Priya and Kritika, this book would not have been come out in the present shape. They had been a driving force in bringing out this publication. While Charu Priya worked on computer to typewrite and set photos, Kritika helped in taking out the shape of the manuscript.

I must acknowledge the contribution of Dr. Neelam Yadav, working as Post-Doctoral Fellow (UGC) with me, for rendering her consistent help and cooperation as and when required, during preparation of the manuscript. Acknowledgements are due to all friends and well wishers who have inspired and encouraged me in my endeavor to finish this book.

Further, I wish to express my thanks to all my research scholars, especially Dr. Jitendra Kumar Mishra, Dr. Indra Dev Chaudhary, Dr. Shyam Kumar Verma, Dr. Pankaj Kumar and Mr. Ram Kewal, for their help and support during the phase of publication of this Book. I would also express my heartfelt thanks to all learned and respected authors whose information I have gathered to complete my book but could not find separate mention. I am thankful to the KALYANI Publishers, for timely publishing the book.

Last but certainly not least, I apologize for any short comings that might have occurred in this publication. I look forward to criticism and constructive suggestions for the improvement of the book.

2020 **NEERJA AGRAWAL**

Contents

Chapter 1

Fruits, Vegetables and Pests

1.1. INTRODUCTION

Fruits and vegetables are important source of nutrition for India's enormous population. India is a major producer of fruits and vegetable after China today, but with increasing globalization; it has become a challenge for this country not only to feed its own population but also to export its produce to various developed countries. For export of mango, restrictions or bans on certification of commodities for export has become difficult. The export of mango is increasing despite concern that fruit fly pests might be shipped along with their fruits. Being unable to use ethylene di-bromide fumigation because of residue problems, the best method at present would appear to be through innovative pest control methods.

India is the second largest producer of fruits in the world with an annual production of 93.71 million MT from an area of 6.46 million hectares and contributes to 14.51 per cent of world's fruit production (Anonymous, 2017). Among various tropical fruits, mango and guava fruits are nutritive and attractive delicious commodities to eat for man. Mango (*Mangifera indica* Linn.) being the most important fruit, nearly 87 countries are known for commercial mango production. China, India, Indonesia, Mexico, Pakistan, Philippines and Thailand are well known as prominent mango producing countries (Tharanathan *et al.*, 2006). Mango described as the "king of fruits", known for its strong aroma, delicious taste, and high nutritive value is a prominent horticulture crop of India. Mango is a tropical and subtropical fruit crop grown in India over an area of 25 lakh hectares with production of 202 lakh tones covering about 38 per cent of area and accounting of 22.6 per cent total production of total fruits in the country, which is highest in the world with India's share of about 54%. It has a high potential of profit and foreign exchange provided quality mangoes are produced. Guava (*Psidium guajava* Linn.) aptly called the apple of tropics is one of the most referred and legendary fruit because of its hardy and positive bearing nature, high vitamin C content and more income with minimum inputs (Singh, 2010). Guava is cultivated on 2.60 lakh hectares of land and its production is about 36.15 lakh tones in India (Anonymous, 2017).

Fruit fly is major limiting factor in production in rainy guava. Infestation of fruit fly ranged 20-46% with annual crop loss of 16-40% in U.P. (Haseeb, 2007) Maggots feed on the pulp of fruit and make it unfit for consumption, therefore, cause 25-50% loss every year (Qureshi *et al.*, 1976).

Both mango and guava are equally attractive to different insect pests. The infestation of insect is a major limiting factor in the production of mango and guava. The vast range of climatic conditions and agro-ecosystems in which mango is grown is subject to the attack of about 500 species of insect pests (Tandon and Varghese, 1985). Out of this, about 400 species have been recorded from Asia and approximately 45 per cent reported from India. Among these 45 per cent, about a dozen are found more severe causing considerable loss to the crop. Large numbers of insect pests have been reported to occur on guava at various growth stages, but a few are a real menace to the cultivation of this crop. More than 80 species of insect pests have been recorded on guava trees affecting the growth and yield (Haseeb, 2007).

With current world emphasis on quality fruit for local consumption and export, insects that blemish fruit by feeding, scratching or ovipositing in the pulp or seed can cause heavy loss. Only fruit flies, seed weevils and lepidopteran larvae actually penetrate the fruit pulp and seed. The feeding of other pests such as *Othris materna* (L.), *Gonodonta pyrgo* (cram), *G. clotilda* (stoll) and *Leptoglossus stigmai* (Herbst.) often extend only into pulp of ripening mangoes.

Most mango producing countries are located in fruit-fly infested areas and producer suffers significant direct and indirect economic losses from fruit fly damage. Fruit flies have a greater impact on international marketing and world trade in agricultural produce.

1.2. FRUITS AND VEGETABLE PRODUCTION

Data from the Union Ministry of Agriculture and Farmers Welfare shows a third consecutive year of drop in production of vegetables and fruits. Estimates for 2018-19 show a 6% fall at 59.8 lakh tones from the 2017-18 production of 63.9 lakh tones. In 2016-17, it stood at 65.6 lakh tones. Fruit production too dropped to 56.1 lakh tones or 1.3% lower than 56.81 lakh tones in 2017-18.

Tamil Nadu is among the largest producer and consumer of fruits and vegetables, but it has done little to help the farmer, reveals the RBI's economic survey. The state has come under severe criticism for lack of agricultural reforms. In the agricultural marketing and farmer friendly reforms index, TN is in the bottom of the pool with a score of less than 20.

Maharashtra topped (Scored 80) in implementation of reforms, most of the marketing that ensured the best environment for doing agri-bussiness across the country. "Almost two thirds of states could not reach the halfway mark of reforms score. Major states like Uttar Pradesh, Punjab, West Bengal, Assam, Jharkhand, TN and Jammu and Kashmir are in this group," the RBI economic survey that released recently noted.

The state horticulture department has been taking baby steps to help farmers by planning to launch an online portal for farmers to sell fruits, to help weed out middlemen and ensure access to good market rates. But till these measures are implemented, it's the farmer and the customer who have to keep paying.

In a year when vegetable and fruit production was low, mango farmers stepped up their output. Mango production in the sate was up by 2.5% at 12.65 lakh tones. "As per the 2nd Advance Estimates of 2018-19, mango production may slightly decraease from 21.82 million tones in 2017-18 to 20.95 million tones in 2018-19. The decrease is estimated primarily due to decline in mango production in the states of Karnataka, West Bengal, Bihar, Telangana and Maharashtra," the minister of agriculture Narendra Singh Tomar said in a Lok Sabha reply.

TABLE 1 : Area and production of fruit crops in India 2016-17
Area in '000 HA Production in '000 MT and Productivity = MT/Ha

Crop	*Area*	*Production*	*Productivity*	*% share*
Banana	852	30275	35.5	35.2
Mango	2267	20295	8.9	23.6
Citrus	1037	12053	11.6	14.01
Papaya	138	6145	44.5	7.1
Guava	260	3615	13.9	4.2
Apple	278	2258	8.1	2.6
Pineapple	114	1969	17.3	2.3
Sapota	105	1252	11.9	1.5
Grapes	136	2784	20.5	3.2
Pomegranate	216	2521	11.7	2.9
Litichi	91	578	6.3	0.7
Others	279	2268	8.1	2.63

TABLE 2 : Area and production of Vegetable in India 2016-17
Area in '000 HA Production in '000 MT and Productivity = MT/Ha

Crop	Area	Production	Productivity	% share
Potato	2151	48237	22.4	30.6
Tomato	799	19542	24.5	12.4
Onion	1293	21718	16.8	13.8
Brinjal	727	12323	16.9	7.8
Tapioca	203	4421	21.8	2.8
Cabbage	394	8720	22.1	5.5
Cauliflower	451	8484	18.8	5.4
Okra	501	5783	11.5	3.7
Peas	540	5252	9.7	3.3
Sweet potato	132	1473	11.1	0.9
Others	1631	21811	13.4	13.8
Total	8822	157764	17.9	

***Source* :** Indian Horticulture Database (2017).

TABLE 3 : Area and production estimates for Horticulture crops DURING 2010-2013
Area in '000 HA Production in '000 MT and Productivity = MT/Ha

	2010-11			2011-12			2012-13		
Crop	Area	Production	PDY	Area	Production	PDY	Area	Production	PDY
Fruits									
Banana	830	29780	35.9	797	28455	35.7	776	26509	34.2
Mango	2297	15188	6.6	2378	16196	6.8	2500	18002	7.2
Citrus	846	7464	8.8	915	7922	8.7	1042	10090	9.7
Papaya	106	4196	39.6	117	4457	38.1	132	5382	40.7
Guava	205	2462	12.0	220	2510	11.4	236	3198	13.6
Apple	289	2891	10.0	322	2203	6.8	312	1915	6.1
Pineapple	89	1415	15.9	102	1500	14.7	105	1571	14.9
Sapota	160	1424	8.9	163	1426	8.7	164	1495	9.1
Grapes	111	1235	11.1	116	2221	19.1	118	2483	21.1
Pomegranate	107	743	6.9	112	772	6.9	113	745	6.6
Litchi	78	497	6.4	80	538	6.7	83	580	7.0
Others	1265	7583	6.0	1383	8224	5.9	1402	9315	6.6
Total	6383	74878	11.7	6705	76424	11.4	6982	81285	11.6
Vegetables									
Potato	1863	42339	22.7	1907	41483	21.8	1992	45344	22.8
Tomato	865	16526	19.1	907	18653	20.6	880	18227	20.7
Onion	1064	15118	14.2	1087	17511	16.1	1052	16813	16.0
Brinjal	680	11896	17.5	692	12634	18.3	722	13444	18.6
Tapioca	221	8076	36.5	227	8747	38.5	207	7237	35.0
Cabbage	369	7949	21.5	390	8412	21.6	372	8534	22.9
Cauliflower	369	6745	18.3	391	7349	18.8	402	7887	19.6
Okra	498	5784	11.6	518	6259	12.1	231	6350	27.5
Peas	370	3517	9.5	408	3745	9.2	421	4006	9.5
Sweet potato	113	1047	9.3	110	1073	9.8	112	1132	10.1
Others	2083	27557	13.2	2352	30459	13.0	2815	33213	11.8
Total	8495	146554	17.3	8989	156325	17.4	9205	162187	17.6

***Source*:** Indian Horticulture Database (2013).

TABLE 4 : Area and production estimates for Horticulture crops during 2013-2016 Area in '000 HA Production in '000 MT and Productivity = MT/Ha

Crop	2013-14			2014-15			2015-16		
	Area	Production	PDY	Area	Production	PDY	Area	Production	PDY
Fruits									
Banana	803	29725	37.0	845	28137	33.2	841	29135	34.6
Mango	2516	18431	7.3	2218	18832	8.5	2209	18643	8.4
Citrus	1078	11147	10.3	923	11742	12.7	1024	11581	11.3
Papaya	133	5639	42.3	122	4682	38.4	132	5667	42.9
Guava	268	3668	13.7	248	3919	15.8	255	4048	15.9
Apple	119	2585	21.8	320	1922	6.0	277	2521	9.1
Pineapple	313	2498	8.0	114	1913	16.8	110	1924	17.5
Sapota	177	1744	9.9	111	1290	11.6	107	1294	12.0
Grapes	110	1737	15.8	122	3205	26.3	122	2590	21.2
Pomegranate	131	1346	10.3	194	2095	10.8	197	2306	11.7
Litichi	84	585	7.0	90	569	6.3	90	559	6.2
Others	1484	9872	6.7	936	7977	8.5	1625	22707	13.9
Total	7216	88977	12.3	6242	86283	13.8	6148	102975	16.7
Vegetables									
Potato	1973	41555	21.1	2069	45951	22.2	2117	43417	20.5
Tomato	1204	19402	16.1	894	19167	21.4	774	18732	24.2
Onion	882	18736	21.2	1150	18736	16.3	1320	20931	15.8
Brinjal	711	13558	19.1	679	12439	18.3	663	12515	18.9
Tapioca	228	8139	35.7	211	5715	27.0	204	4344	21.3
Cabbage	400	9039	22.6	380	8795	23.1	394	8806	22.3
Cauliflower	434	8573	19.8	414	8026	19.4	426	8090	18.9
Okra	533	6346	11.9	493	5552	11.3	511	5849	11.4
Peas	434	3869	8.9	467	4476	9.6	498	4811	9.7
Sweet potato	106	1088	10.3	111	1336	12.0	126	1454	11.5
Others	2492	32591	13.1	2623	36838	14.0	1625	22707	13.9
Total	9396	162897	17.3	9494	167030	17.5	8658	151656	17.5

Source : Horticulture Statistics Division, Department of agriculture, Cooperation and Farmers welfare (2016)

Mango is an important fruit of India. It is known as king of fruits which is grown from foot hills of Himalayas to Cape Comarin and suites the Indian climate due to its Sub-tropical climate. It is available to the consumer right from mid March to mid September for one region or the other, longest any fruit available in fresh form in our country. The earlier varieties are from south and the availability extends to April-July to September northwards.

Although mango is now being grown in more than 87 counties of the world but no where it is so greatly valued as in India where it occupies an area 2267 million hectares with a total production of 20295 million metric tones, contributing 40% of the area devoted to food crops in India. India contributes a major share (70%) in world production of mango, which is in the order of 13.44 million metric tones which is too much of production as compared to other countries. The major mango growing states are Uttar Pradesh, Bihar, Andra Pradesh, Gujrat, Maharshtra and Tamil Nadu.

Mango is grown under highly diversified climatic conditions and is subjected to attack by 230 pests in India. More than 492 species of insects, 17 species of mites and 26 species of nematodes have been reported to be infesting mango trees, about 45% of which have been reported from India. Almost a dozens of them have been found damaging the crop to a considerable extent causing severe losses and, therefore, may be termed as major pest of mango. These are hopper, *Idioscopus clypealis* Lethierry, *Idioscopus nitidulus* (Walker) and *Amritodus atkinsoni* Lethierry; mealy bug, *Drosicha mangiferae* Green; inflorescence midge, *Erosomyia indcia* Grover; fruit fly, *Bactocera (Dacus) dorsalis, B. zonata, B. correcta* etc.; scale insects, shoot borer, leaf webber, stone weevil and pulp weevil. Of these, insects infesting the crop during flowering and fruiting period cause more severe damage and are termed as major pests of mango.

The perennial nature of crop, large scale monoculture, high density planting, high yielding activation, and abuse and misuse of fertilizers and pesticides, the mango eco-system is completely disrupted. This has led to ecological crises and serious consciousness in terms of pest problems. The insect pest problems are categorized in three groups :

- Key pests of National level,
- Region specific pests,
- Minor but potential pests.

Fruit fly is considered as a major constrain in export of mango fruits. Fruit flies are economically important pests of many fruit crops and vegetables. Out of 4,400 known species of fruits flies, only 200 species are found infesting crops in various parts of India. Fruit flies infestation results in severe loss of the crop. The damage may extend from 30 to 35% and some times may go up to a maximum of 100%. The annual losses due to fruit flies are estimated as ₹ 2,945 crores in fruit crops *viz.* mango, guava and chiku.

1.3. FRUIT FLIES AS PEST OF HORTICULTURAL CROPS

Fruit flies are important pests of several fruits and vegetable crops throughout the tropical and subtropical world. Nearly 35% of the known species attack soft fruit of Guava, mango, ber, peach, citrus and several Cucurbitaceous vegetables are important. Fruit flies belong to family Tephritidae which is one of the largest, most diversified and fascinating acalypterate family of order Diptra. Tephritidae include about 400 species arranged in 500 genera (White and Harris, 1992). These are commonly called Fruit flies due to close association with fruits and vegetables. Out of 400 species, 392 species have been recorded in India, (Kapoor, 1993). Fruit flies are of great economic importance as majority of them cause extensive damage to many fruits and vegetable crops. Female fly lays eggs in soft skin of fruit and ruins more than 400 different fruits and vegetables including Guava, Mango, Plum, Citrus, Melon, Papaya, Peach, Passion Fruit, and Apple. They are considered as high priority quarantine pests. India has been included in the list of those countries from where fruit import to developed countries is banned.

Fruit flies (Diptera : *Tephritidae*) include some of the world's most serious pests. Besides causing direct loss of millions of rupees, to a wide variety of fruits, vegetables and flower crops, they limit the development of agriculture in many countries, because of the strict trade quarantine imposed to prevent their spread. Of nearly 5000 species known worldwide, categorized in almost 500 genera, nearly 200 are considered pests. Many species are extremely similar and therefore their identification is very difficult. Description, recognition, and genetic analysis are constantly changing the taxonomy of fruit fly family. Tephritidae is one of two families referred as "fruit flies". Tephritidae does not include the biological model organism of genus *Drosophila*, which is often called the "common fruit fly". *Drosophila* is, instead the type genus of the second "fruit fly" of family Drosophilidae. Tephritid fruit flies are of major importance in agriculture. Some have negative effects, some positive. Various species of fruit fly cause damage to fruit and other plant crops. The genus *Bactrocera* is of worldwide notoriety for its destructive impact on agriculture. Some fruit flies are used as agents of biological control, thereby reducing the population of pest species. Several species of fruit fly genus *Urophora* have been shown to be effective agents against rangeland destroying weeds. Most fruit flies lay their eggs in plant tissues, where the larvae find their first food upon emerging. The adults usually have a very short lifespan. Some live for less than a week.

Most of fruit flies are polyphagous pest having wide range of host plants. The female flies puncture the soft and tender fruit with their stout and hard ovipositor and lay egg below the epidermis at the rate of 4 to 10 eggs per fruit each time. *Bactrocera* spp. are pests of major importance in the eastern hemisphere. *Bactrocera* female fruit fly inserts its eggs beneath the skin of fruit specially at ripening stage. About 1200-1500 eggs are laid by *B. dorsalis*, a major pest of mango. The larvae tunnel into the fruit, contaminating it with frass and providing entry for fungi and bacteria. When the infected fruit is immature, it ripens prematurely and is unfit for marketing. Fully grown larvae measuring 7 mm drops to the ground and enter the soil, where they pupate. The egg stage lasts from 2-20 days. There are usually three larval instars. The larval and pupal periods are each 2-4 weeks. After emergence, the females require a protein source for egg maturation. Studies conducted by Singh (1991) with *B. dorsalis* in India indicated that pupal period was longest at 15°C and shortest at 35°C. Warm humid weather is considered to be favourable for *Bactrocera* fruit flies and pest population build up as mango ripening occurs; and its population decrease during dry periods. Up to 30% of mango fruits were attacked by *B. dorsalis* in July and August in central Punjab, maximum being in August with 35% of fruits damaged by *B. dorsalis* and *B. zonata* (Mohyuddin and Mahmood 1993). Singh

(1991) indicated that the frequency of *Bactrocera* injury to physiologically mature fruits *Dashehari*, ranged from 3.6% to 10% whereas in fully ripe fruits the figure ranged from 10% to 25.9%. Highest damage was reported in fully ripe fruit of "*Mallika*" followed by "*Totapari*" (Pene *et al.*, 1998).

It is difficult to assess the losses caused by fruit flies but as quoted by Butani (1987) more than 50 per cent of fruits are either partially or fully damaged by one or the other species of fruit fly. The damage caused is very severe in certain cultivated crops. *Bactrocera (Dacus) dorsalis* and *B. cucurbitae* caused heavy damage to fruits and vegetables (Agrawal and Mathur 1988). The adults of *Bactrocera* spp. are strong flyers and travel 3-40 km in search of host fruits. A preliminary survey reports in India indicate that overall losses without control may be about 25 per cent in cucumber, 30 per cent in gourds and 40 per cent in melon.

In guava, Gowda (2018) reported that samples of total fruits and damaged fruits were collected every week and their mean was recorded on monthly basis from November 2017-April 2018 and percentage of every month was calculated. It was observed that samples collected from first farmer, showed that maximum fruit fly infestation was recorded during April (47.60%). However, the average percentage of damaged fruits was recorded as approximately 37.44% in guava orchard, which was maximum among all the farmers. In case of samples collected from second farmer the damage was approximately 27.14%. In the third farmers field, approximately 30.39% damage of fruits was recorded during the period of survey. Minimum percentage of damage of guava fruit was recorded as 18.40% due to fruit flies in the survey conducted in Kanpur, Central Uttar Pradesh in Guava orchards. It is worth mentioning here that the infestation was low during November 2017–January 2018 as the fruit was hard but after that fruit became more prone to fruit fly infestation due to ripening.

As regards species composition of fruit flies infesting different varieties of guava and mango, only two species, *B. dorsalis* and *B. zonata* were identified during laboratory experiments by Kumar (2018). Among the five varieties of guava, Allahabad Safeda was highly infested with both the species followed by Lucknow 49 and Seedless. Least infestation was found in rough skinned variety of guava Pear shape. In mango, variety Deshahari was proved harbouring maximum number of fruit flies while, variety Safeda registered minimum counts of fruit fly emerging from this mango variety. It is worth mentioning here that guava varieties recorded more infestation by *B. zonata* as compared to *B. dorsalis*. In mango, however, *B. dorsalis* out numbered *B. zonata*.

❑❑❑

2

Chapter

Important Horticultural Crops

2.1. MANGO

Mangoes are tropical fruits that are not only juicy, flavorful, and pretty to look at but are also good for your health. They're high in vitamins, minerals, and antioxidants. Years ago, they were hard to find and considered exotic, but these days mangoes are readily available year round. *Mangifera indica* (MI), also known as mango, has been an important herb in the Ayurvedic and indigenous medical systems for over 4000 years. Mangoes belong to genus *Mangifera* which consists of about 30 species of tropical fruiting trees in the flowering plant family Anacardiaceae. According to ayurveda, varied medicinal properties are attributed to different parts of mango tree.

Mango is one of the most popular of all tropical fruits. Mangiferin, being a polyphenolic antioxidant and a glucosylxanthone, it has strong antioxidant, anti lipid peroxidation, immunomodulation, cardiotonic, hypotensive, wound healing, anti-degenerative and antidiabetic activities.

Ripe mango fruit is considered to be invigorating and freshening. The juice is restorative tonic and used in heat stroke. The seeds are used in asthma and as an astringent. Fumes from the burning leaves are inhaled for relief from hiccups and affections of the throat. The bark is astringent, it is used in diphtheria and rheumatism, and it is believed to possess a tonic action on mucus membrane. The gum is used in dressings for cracked feet and for scabies. It is also considered anti-syphilitic. The kernels are converted into flour after soaking in water and eliminating the astringent principles. Most parts of the tree are used medicinally and the bark also contains tannins, which are used for the purpose of dyeing.

One cup of mango pieces has 99 calories, almost no fat, and 25 grams of carbohydrates. It is also almost sodium free and has almost 3 grams of fiber. Mangoes are high in vitamins, potassium, and folate and also add fiber to your diet. Fiber is essential for a healthy digestive system and helps keep you feeling full between meals. It also slows down the absorption of sugar after you eat.

Mango Nutrition Facts :

Serving Size 1 Cup Pieces

Per Serving	**% Daily Value***
Calories 99	
Calories from Fat 5.7	
Total Fat 0.6 g	1%
Saturated Fat 0.1 g	1%
Polyunsaturated Fat 0 g	
Monounsaturated Fat 0 g	
Sodium 1.7 mg	0%
Potassium 277 mg	8%
Carbohydrates 25 g	8%
Dietary Fiber 2.6 g	10%
Sugars 23 g	
Protein 1.4 g	
Vitamin A	36%
Vitamin C	100%
Calcium	1.8%
Iron	1.5%

Health Benefits

Mangoes are high in vitamin C. In fact, one mango has all the vitamin C you need for a whole day. Vitamin C is essential for immune system function, strong connective tissue, and healthy blood vessel walls. Getting an insufficient amount of vitamin C every day can lead to bruising and make it difficult for wounds to heal properly.

Mango is also high in potassium and has almost no sodium, so eating mango may help regulate blood pressure and body fluid balance. Mangoes also boast high levels of folate and vitamin A.

Folate is a B-complex vitamin that is important for heart health and production of blood cells. Vitamin A is needed for normal vision, healthy skin, reproductive health, and normal cell development.

Mangoes also contain quercetin, mangiferin, and norathyriol, which are all potential antioxidant compounds. Antioxidants help protect our body's cells from damage from free radicals (which experts believe cause cancer, atherosclerosis, and other diseases).

Although not many people choose to eat mango skin, it is edible. If you want to give it a try, depending on the mango, you might find that it tastes slightly bitter, chewy, and possibly tough. On the nutrition side, however, it's filled with various antioxidants and fiber. Note that it also contains urushiol, the same compound that causes reactions to poison ivy, so if you're sensitive be cautious about consuming mango skin.

Mango is low in calories but full of nutrients.

One cup (165 grams) of sliced mango provides :

- **Calories** : 99
- **Protein** : 1.4 grams
- **Carbs** : 24.7 grams
- **Fat** : 0.6 grams
- **Dietary fiber** : 2.6 grams
- **Vitamin C** : 67% of the Reference Daily Intake (RDI)
- **Copper** : 20% of the RDI
- **Folate** : 18% of the RDI
- **Vitamin B6** : 11.6% of the RDI
- **Vitamin A** : 10% of the RDI
- **Vitamin E** : 9.7% of the RDI
- **Vitamin B5** : 6.5% of the RDI
- **Vitamin K** : 6% of the RDI
- **Niacin** : 7% of the RDI
- **Potassium** : 6% of the RDI
- **Riboflavin** : 5% of the RDI
- **Manganese** : 4.5% of the RDI
- **Thiamine** : 4% of the RDI
- **Magnesium** : 4% of the RDI

It also contains small amounts of phosphorus, pantothenic acid, calcium, selenium and iron.

2.1.1. Insect Pests of mango

There are number of insect pests of this fruit and over 175 species of insect have been reported damaging mango tree but the most abundant and destructive at the flowering stage are the mango hoppers. Also mango mealy bug in North India, stem borer, fruit fly, mango nut weevil and caterpillar pests play a major role in bringing down the yield. It is almost necessary to control these pests otherwise there is a heavy fruit drop and the trees may remain without fruit.

Major pests

Common name	Scientific name	Family	Order
Mango hoppers	*Idioscopus niveosparsus* Lethierry, *I. clypealis* Lethierry, *Amritodus atkinsoni* Lethierry	Cicadellidae	Homoptera
Mango mealy bug	*Drosicha tnangiferae* Green	Margarodidae	Hemiptera
Fruit fly	*Bactrocera dorsalis* Hendel	Tephritidae	Diptera
Mango nut weevil	*Sternochaetus tnangiferae* Fabricius	Curculionidae	Coleoptera
Stem borer	*Batocera rufomaculata* DeGeer	Cerambycidae	Coleoptera
Mango leaf webber	*Orthaga eiiadrusalis* Hampson	Noctuidae	Lepidoptera
Shootgall psylla	*Apsylla cistellata* Buckton	Psyllidae	Homoptera
Inflorescence gall midge	*Procystiphora tnangiferae* Felt, *Dasineura amaramanjarae* Grover, *Erosomyia indica* Grover	Cecidomyiidae	Diptera

Other pests associated with mango

Common name	Scientific name	Family	Order
Bark eating caterpillar	*Indarbela tetraonis, I. quadrinotata*	Metarbelidae	Lepidoptera
Shoot borer	*Clumetia transversa*	Noctuidae	Lepidoptera
Leaf caterpillar	*Bombolelia jacosairix*	Noctuidae	Lepidoptefa
Leaf miner	*Acrocercops syngramma*	Gracillariidae	Lepidoptera
Leaf twisting weevil	*Apoderns transquebarius*	Curculionidae	Coleoptera
Red ant	*Oecophylla smaragdina*	Formicidae	Hymenoptera
Eriophyid mite	*Aceria tnangiferae*	Eriophyidae	Acarina
Fruit sucking moths	*Eadocima (Othre is) fullonia*	Noctuidae	Lepidoptera
Aphids	*Toxoptera odinae*	Aphididae	Homoptera
Fruit borers	*Deanolis albizonalis*	Crambidae	Lepidoptera
Termites	*Odoniotermes* sp.	Termitidae	Isoptera
Leaf cutting weevil	*Apoderus traansquebarius*	Curculionidae	Coleoptera
Thrips	*Selenothrips rubrucinctus*	Thripidae	Thysanoptera

Mango hoppers (*Cicadellidae* : Homoptera)

Three species of hoppers are found to feed on the inflorescence of mango *Amritodus atkinsoni* (Lethierry), *Idioscopus clypealis* (Lethierry) and *I. niveosparsus* (Lethierry).

Distribution : It is most serious of all the mango pests and prevalent all over the country causing heavy damage to mango crop. In India, these hoppers are widely distributed in all the mango growing regions. *A. atkinsoni* is more common in North India. *I. clypealis* is found all over India, more predominant in South Gujarat, Maharashtra and Karnataka. *I. niveosparsus* is recorded from peninsular India.

Host range : Mango.

Nature of damage : Leaf hoppers' population on inflorescence has significant negative correlation with fruit set. The nymphs and adult insects suck the sap from the inflorescence in large numbers causing withering and shedding of flower buds and flowers which result in heavy loss ranging from 25-60 per cent due to poor fruit setting. The honey dew excreted by them affords conditions for development of sooty mould. Egg laying also inflicts injury to the inflorescence.

Life history : Though hopper population exists throughout the year in mango orchards but occasionally it advances during January to April on flowering flush, also noticed during June-August on vegetative flush. Old, neglected and closely planted orchards that are shady and with high humid conditions favour their multiplication.The mango leaf hoppers are small wedge shaped insects and measure 3.4-5.1 mm in length. *I. niveosparsus* is slightly smaller with three spots on the scutellum and prominent white band across its light brown wings. *I. clypealis* is the smallest with two spots on the scutellum and dark spots on the vertex and is light brown in colour. *A. atkinsoni* is the largest and light brown having two spots on the scutellum. The female hopper inserts the eggs into flower buds and the inflorescence stalk. The nymphs hatch out in 4 to 7 days. Freshly hatched nymphs are wedge shaped and whitish in colour with two small red eyes. Gradually with each moulting, the colour changes to yellow, yellowish green, green and ultimately greenish brown. The period from egg to adult takes about 12-17 days and during a flowering season two or more broods of the pest may occur.

Natural enemies : Three species of coccinellid beetles, namely, *Coccinella septempunctata, C. transversalis* and *Cheilomenes sexmaculata* are most common predators on nymphs of leaf hopper. Besides these, spiders-*Araneus sinhagadensis, Cyrtophora cicatrosa, Cheiracanthium danieli,* etc. *Stagodyphus sarsinorum, Harsilia savigryi, Oxyopes shweta, Rhene indicus, Phidippus* sp. also keeps hopper population in check. The reduviid, *Endochus inornalus* has been observed in South India predating upon nymphs of *Idioscopus niveosparsus.* Entomopathogenic fungi, *viz., Verticillium lecanii, Beauveria bassiana, Hirsutella* spp., *Isaria* sp. *Pleospora infectoria* and *Cladosporium* sp. are also effective in areas of high humidity.

Management strategies :

- Avoid dense plantings, maintain open canopy, prune overcrowded overlapping branches after rainy season.
- Orchards should be kept clean by regular ploughing and removal of weeds.

- Protecting and encouraging biocontrol agents like predator, *Mallada boninensis, Chrysopa lacciperda,* egg parasite, *Polynema* spp., *Gonatocerus* sp., *Tetrastichus* sp. and fungus, *Verticillium lecanii.*
- Spraying of 0.2% Nimbicidin or Azadirachtin 3000 ppm @ 2.0 ml/1 at initial stage of hopper population.
- Application of bio-agents, *Metarhizium anisopliae* @ 1×10^8 cfu/ml or *Beauveria bassiana* @ 10 cfu/ml on tree trunk once during off season and twice at 7 days interval during flowering season.
- Spray Lambda cyhalothrin 5% EC @ 0.5 ml or imidacloprid 200 SL @ 0.25 ml/l or Thiamethaxam (0.05%) or propanophos (0.05%). First spray should be done at early stage of panicle formation if hopper population, is more than 5-10 panicle, second spray at full length stage of panicle and the third spray after fruit setting (at pea size).
- Chemical spray is to be minimized and should be need based.
- A rational rotation of insecticide is desirable to counteract the tendency of pest to develop field resistance.

Mango mealy-bug, *Drosicha mangiferae* (Green) (Margarodidae : Hemiptera)

Distribution : More than 20 species of mealybugs are known to attack mango (Tandon and Verghese 1985). However, three are serious pests-*Drosicha mangiferae, Drosicha stebbingi* and *Rastrococcus iceryoides.* Of these, the first two species are present in India, Nepal, Bhutan, China, Pakistan and Bangladesh, while *Rastrococcus iceryoides* is reported from Malaysia. In India, it is widely distributed along the Indo-gangetic plains and found in Punjab, Uttar Pradesh, Bihar and Delhi causing severe damage to mango crop.

Host range : These are polyphagous and infest more than 70 plant species (Tandon and Lal 1978). It is reported to attack a number of fruit crops such as apple, apricot, ber, cherry, citrus, falsa, fig. grapevine, guava, gular, jack-fruit, jamun, litchi, mulberry, papaya, peach, plum and pomegranate.

Nature of damage : Nymphs and female bugs suck sap from inflorescence, tender leaves, shoots and fruit peduncles. Affected panicles shrivel and get dried. Severe infestation affects the fruit set and causes fruit drop. When they are in large number they devitalize the plant and they produce honeydew which encourages growth of sooty mould, giving a very unhealthy look to the plant as a whole. At times, they are found clustering in masses on young shoots, like fungus outgrowths.

Life history : There is a well-established sexual dimorphism in the adult stage which is generally found during the mid summer period, *i.e.* from April to June. Adult females are wingless and large-bodied. The male is a winged creature with only one pair of wings and a very delicate reddish body which flies actively and fertilizes the females. The adult gravid females after fertilization crawl down along the tree-trunk to the ground where they lay eggs at depths of about 5-15 cm and in clusters of 300-400 eggs each. The oviposition is generally confined to an area near and around the base of the tree. These activities of migration from the tree downwards to the ground and oviposition in the soil are generally confined to the months of April, May and June. The males die soon after mating and the females soon after oviposition. The eggs laid in the soil take quite a few months before they hatch and is influenced by the temperature and moisture conditions of the soil. Hatching can be as early as November of the same year or as late as March of the succeeding year. The young nymphs soon after hatching crawl about in search of some suitable food-plant on which, if found, they spend some time. Thereafter, they begin their ascent along the tree-trunks and this upward migration lasts for several weeks. The late instar nymphs and adult female mealybugs are flat, oval and waxy white. They moult thrice during their nymphal period which lasts about three months or more, depending on the environmental temperature. Thereafter, the nymphs developing into males undergo some sort of pupation and transform themselves into winged adults and the female-producing nymphs do not undergo any appreciable change except in size. Thus, there is only one generation during the year. Unlike many other coccids, the nymphs of this pest do not remain stationary although they are sluggish.

Natural enemies : Several species of natural enemies have been observed predating upon nymphs of *D. mangiferae* and *D. stebbingi.* These include 11 species of coccinellids, six species of spiders, two species of mites and three species of parasitoids (Tandon, 1995).

Management strategies :

- Proper orchard maintenance by removal of weeds that harbor mealy bugs.
- Ploughing of the orchard during November-December.
- Flooding of orchard with water and raking of soil around tree trunk exposes the eggs to sun and natural enemies thereby destroying them.
- After mud plastering, banding of tree trunk with alkathene (400 gauge), 25 cm wide sheets should be fastened to the tree trunk with the help of sutli, 30 cm above ground level one week before their emergence, *i.e.*, around November/December.
- Spray NSKE (4%) or garlic oil (1%) on tree trunk below the alkathene band to kill the crawlers congregated (Tandon 1995).
- Conservation of bio control agents, *Beauveria bassiana,* predators, *Menochilus sexmaculatus, Rodolia fumida* and *Sumnius renardi* by avoiding use of broad spectrum insecticide during their peak activity period and minimum tillage.

- Application of *Beauveria bassiana* product (2 g/litre or 10×10^7 spores/ml) in last week of January around tree trunk gave complete mortality of nymphs within 5-12 days.
- Release of predators, *Menochilus sexmaculatus, Rodolia fumida, Sumnius renardi* and Australian ladybird beetle, *Cryptolaemus montrouzieri* @ 10-15 No./tree are effective in controlling the nymphs of the mealy bug.
- Application of Methyl Parathion dust 2% @ 250 g per tree in the soil around the trunk during 3rd or 4th week of December.
- Early instar nymphs of the mealy bug can be controlled by spraying of 0.05% carbaryl or Dimethoate (0.04%) from January to March.

Fruit fly, *Bactrocera dorsalis* Hendel (Tephritidae : Diptera)

Distribution : Fifty two species are placed in the *Bactrocera dorsalis* complex in Asia and eight of which are considered of economic importance (Drew and Hancock 1994). The Oriental fruitfly, *B. dorsalis* has been reported from India, Sri Lanka, Myanmar, Nepal, Bhutan, Taiwan, Thailand, Laos, Vietnam and Cambodia in Asia. It is widely distributed in the Orient region from Australia and Hawaii to Pakistan, hence it is also called Oriental fruit fly. The pest is active throughout the year in South India whereas in northern parts the pest hibernates during winter (November to March) in pupal stage.

Host range : This is a polyphagus species in nature and extent of damage depends upon type, size and condition of fruit, availability of different hosts and population density. The most preferred hosts in India are mango, guava, litchi, citrus, papaya, custard apple, etc.

Nature of damage : Attacked fruits usually show signs of oviposition punctures and ripe fruits with high sugar content exude a sugary liquid. On hatching the maggots feed on pulp of those fruits. As a result a brown patch appears around the place of oviposition and the infested fruits start rotting. These affected fruits drop down prematurely and the maggots come out from these fallen fruits to pupate in the soil. Semi ripe fruits are attacked usually by April-May. Sometimes it becomes serious.

Life history : The flies appear late in spring on such of the fruits that are about to ripe and the population increases rapidly during summer (May-July). The adult fly is light brown with transparent wings. Adult flies are very conspicuous. These are about 7 mm long, with hyaline wings (expanse : 13-15 mm), thorax ferrugineous without yellow middle stripe, legs yellow, abdomen conical in shape and dark brown in colour. Preoviposition period is 2-5 days. The female flies lay eggs just below the fruit epidermis (1-4 mm deep). A single female can lay 150-200 eggs (average 50) in about a month. The eggs are laid in clusters of 2-15 eggs and these hatch in 2-3 days during March and 1-1 54 days during April. Maggot duration is 6 days in summer and extends up to 19 days with the fall in temperature. Pupation usually takes place 80-160 mm below the soil surface and pupal period ranges from 6 days (summer)-44 days (winter).

Natural enemies : Eighteen species of parasitoids (mostly *Opius* spp.) have been reported parasitizing *B. dorsalis* in India (Tandon 1995). However, recent surveys indicate that many parasitoids reported earlier are not available.

Management strategies :

- Prior to harvest (30-40 days) collect and dispose off infested and fallen fruits to prevent further, multiplication and carry over of population. Do not retain birds' damaged fruits on trees because they attract adult fruit flies.
- Ploughing of orchard during November-December to expose pupae to sun's heat which kills them.
- For conservation of parasitoids, collect all fallen infested fruits and put in a drum covered with fine wire mesh. This will allow the parasitoids to emerge while fruitfly will remain in the drum.
- If infestation is heavy, bait splash on the trunk only, once or twice at weekly interval is recommended. To prepare bait splash, mix 100 gm of jaggery in one litre of water and add 1 ml of Deltamethrin by using an old broom.
- To control adult flies during severe infestation placing poison bait *viz.* Protein hydrolysate + malathion 50 ml + 200 ml molasses in 2 litres of water be sprayed adding an additional 18 liters of water to bait poison. Commencing at pre oviposition period and repeat at 15 days interval. Addition of 10 ml methyl eugenol in place of molasses is also recommended.
- Three weeks before harvesting, spray Deltamethrin 2.8 EC @ 0.5 ml/I + Azadiractin (3000 ppm) or @ 2 ml/1.
- Harvest one week early, the fully mature fruits from trees to escape oviposition.
- Post-harvest immersion of mango fruits in hot water at 48°C for 45-60 minutes gave 100% mortality of fruitfly eggs in the epicarp without affecting fruit quality (Jothi *et al.,* 1994).
- Irradiation of fruits 400 G-rays using cobalt 60 to control fruit fly.
- Managing fruit flies also reduces anthracnose disease and prevents late fruit fall.

Mango nut weevil or stone weevil *Sternochetus mangiferae* (Fb.) (Curculionidae : Coleoptera)

Distribution : Three species of stone weevil, namely, *Sternochetus mangiferae, S. gravis* and *S. frigidus* have been reported as pests of mango (Tandon and Verghese 1985). *Sternochetus mangiferae* is widely distributed in Asia covering Bangladesh, Cambodia, India, Indonesia, Japan, Malaysia, Myanmar, Philippines, Pakistan, Sri Lanka, Thailand and Vietnam. The other species, *S. gravis,* is found in India, Indonesia, Malaysia,

Myanmar and Pakistan while *S. frigidus* is restricted to Malaysia, Pakistan and Thailand. It is widely distributed in the tropics. The pest is more common in South India where late varieties suffer the most.

Host range : Mango

Nature of damage : Eggs are laid singly on the epicarp of partially developed fruits or under the rind of ripening fruits. The grubs as soon as they hatch out from the eggs tunnel in a zigzag manner through the pulp, endocarp and the seed coat and finally reach the cotyledons. As the fruit develops the tunnels get closed up. The grubs feed on the cotyledons and destroy them. The adults who emerge from the pupae also feed on the developing seed and this may hasten the maturity of infested fruits. The adults hibernate in between the crevices on the tree trunks. The weevil attacks only mango.

Life history : The dark brownish stout weevil measures about 6 mm long. The female scoop out the surface of the developing fruit (till it is half ripe) and deposits the eggs singly. On a fruit 12-36 eggs may be deposited. However, finally only a maximum of about 7 weevils can be noticed in a highly susceptible variety. The fluid that oozes from the fruit covers the egg. The incubation period is 7 days. The grub is apodous, fleshy, light yellow with a dark head and pupates inside the nut itself. It emerges as adult in 7 days. The total life cycle from egg to adult occupies 40-50 days. The weevils hibernate from July-August till next fruiting season. There is only one generation in a year.

Natural enemies : Nuclear polyhedrosis virus was recorded from Bangalore on grubs of S. *mangiferae* (Shukla *et al.,* 1984). *Sternochetus gravis* was preyed upon by ants – *Oecophylla smaragdina, Camponotus* sp., *Monomorium* sp., fungus *Aspergillus* sp. and mite *Rhizoglyphus* sp. developed on larvae pupae and hibernating adults in Tripura (De and Pandey 1988).

Management strategies :

- Collection and destruction of infested and fallen fruits at weekly interval till harvest of fruit.
- Ploughing of orchard after harvest to expose hibernating adults, reduce infestation levels.
- Destroy all left over seeds in the orchard and also in the processing industries.
- Vapour heat treatment of fruits.
- Sticky bands should be applied at upper end of tree trunk to prevent migration of weevils to branches for egg laying on fruits during February.
- Spray main trunk, primary branches and junction of branches prior to flowering (November, December) with carbaryl (0.2%) or fenthion (0.1%) or chlopyriphos 20 EC @ 2.5 ml/l to control beetles hiding in the bark.

- Spraying Dimethoate (0.1%) twice at 15 days interval when fruits are of marble size.
- Spray Acephate 75 SP @ 1.5 g/l when fruits are of lime size (2.5-4 cm diameter) followed by Deltamethrin 28 EC @ 1 ml/l after two or three weeks.

Mango stem borer, *Batocera rufomaculata* Deg. (Cerambycidae : Coleoptera)

Distribution : Widely distributed all over India and Bangladesh.

Host range : It attacks apart from mango other trees such as fig, rubber, jackfruit, eucalyptus, mulberry etc. The mango varieties Amlet and Mulgoa are highly susceptible to the attack of this insect.

Nature of damage : The grubs feed by tunneling through the bark of branches and cause wilting. Though it is an occasional pest of importance, in case of severe attack the trees succumb. Normally, the attack goes unnoticed till the branch start drying up. Sometimes sap and frass may be seen exuding from the bore holes.

Life history : Beetle is dark with a fine grayish vestiture and 2 kidney-shaped orange yellow spots on pronotum. The eggs laid singly on the bark or in crevices on tree trunk or branches hatch in about 1-2 weeks. Full grown grubs are cream coloured, fleshy and apodous with dark brown head and measure 90 × 20 mm in size.The grub feeds for 3-6 months and pupates inside the tunnel itself. The adult emerges in about 4-9 months. Beetle emerges in July-August. There is only one generation of the pest in a year.

Management strategies :

- Keep orchard clean and healthy.
- Cut and destroy affected branches with grubs and pupae.
- Clean hole and insert cotton wool soaked in emulsion of dichlorvos (0.05%) or kerosene or petrol in each hole and plug them with mud.
- Use of green muscardine fungi, *Metarhizium anisopliae* or *Beauveria bassianna.*

Mango leaf webber, *Orthaga euadrusalis* Hampson (Noctuidae : Lepidoptera)

Distribution : Hampson (1896) recorded *Orthaga euadrusalis* from India, Sri Lanka and Indonesia, and this has been reported also from Thailand and Philippines. It is a pest that is attaining serious proportions mainly in North India especially in old, crowded orchards where there is excessive shade.

Host range : Mango

Nature of damage : The larvae web the leaves and the terminal shoots into clusters. A webbed cluster of leaves harbour several larvae in the initial stage. The larvae are gregarious in the beginning and feed by scrapping the leaf surface. Late instar larvae feed individually on whole leaf lamina leaving only midrib. As a consequence of severe feeding, clusters of webbed leaves become dry and colour changes to brown. In case of severe infestation, photosynthetic activity is severely hampered. Affected trees, present sickly appearance and can be observed from a distance due to brown dry clustered leaves. Though precise data on crop losses is lacking, Singh (1980) reported that 90% of the completely defoliated shoots dry and fail to fruit in the forthcoming season.

Life history : Moth is grayish with brownish wings and has wavy lines on fore wings. Adults lay upto 30-50 yellowish green eggs singly near the leaf veins. Egg period is 4 days. Caterpillar pale greenish with brown head and prothoracic shield. Pupation takes place inside the webs in silken cocoons. Adult emerges in 11-14 days. The last generation pupates in the soil around December-January.

Natural enemies : Srivastava and Tandon (1980) recorded four species of larval parasitoids and three species of pathogens. *Tetrastichus Israeli,* a pupal parasitoid was recorded from South India. These natural enemies, if conserved and augmented, can play important role in biological control of *Orthaga* spp. in South India.

Management strategies :

- Inspect your mango orchards at least once in a month and search for webbed leaves/shoots.
- Systematically prune the infested clusters and destroy them along with larvae of *Orthaga* spp. Also collect and destroy the webbed infested fallen leaves on the ground under infested trees harbouring old larvae/pupae.
- In case of old dense orchards, prune some side branches and open the tree canopy from centre to enable proper sunlight penetration.
- Encourage the activity of predators, carabid beetle *Panera lacticincta* and reduvid *Oecama* sp. The release of egg parasitoids such as *Trichogramma chilonis* or *T. pretiosum* may be effective.
- Spray the affected trees after pruning with carbaryl (0.1%) or quinalphos (0.05%) or endosulfan (0.05%). Repeat spray at 10-12 days interval. The use of same chemical for every spray should be avoided.

Shoot gall psylla, *Apsylla cistellata* Buckton (Psyllidae : Homoptera)

Distribution : The psyllid has been observed on mango in plains of northern India, Nepal and Bangladesh (Tandon and Verghese 1985). It is a serious pest of North India and particularly reported from Uttar Pradesh, Bihar and Terrai regions of northern India.

Host range : Mango

Nature of damage : Nymphs emerge during August September and suck cell sap from adjacent buds. As a result of feeding, buds develop into hard conical green galls, more than 70% of which get dried (Singh 1980). The galls are usually seen during September-October. Consequently there is no flowering and fruit setting. Severely infested tree yields only 10% of the actual yield.

Life history : Adult females lay eggs into the midrib of leaves in March-April. Freshly laid eggs are white and hatch in the last week of August. Nymphs are flat-oval in shape and yellowish in colour but change in size and colour with time, translucent with their tips partially exposed. Five nymphal instars are present and nymphal period is 140 days. Second instar nymph migrates to the already formed gall. Nymphs pass winter inside the galls. Adult psyllids are black in colour and measure 1.72 mm (male) and 2.50 mm (female). Adults may live up to 30-72 h. It has a single generation per year.

Management Strategies :

- New mango orchard in humid region need to be discouraged.
- Pruning of galls bearing shoots is an affective measure to reduce infestation during forthcoming year.
- Galls with nymphs should be collected and destroyed.
- Quinalphos (0.05%) or dimethoate (0.06%) or monocrotophos (0.05%) were found quite effective if sprayed thrice at 15 days interval during vegetative bud burst stage (Srivastava *et al.,* 1982).
- Spray 2, 4-D (150 ppm, *i.e.* 150 mg/liter of water) during October which opens the galls and nymphs come out and are killed with cold.

Inflorescence gall midges, *Procystiphora mangiferae* Felt, *Dasineura amaramanjarae* Grover, *Erosomyia indica* Grover (Cecidomyiidae : Diptera)

Distribution : Three species of inflorescence gall midges are found throughout India. It is becoming serious in some pockets of Uttar Pradesh, as well as Maharashtra.

Host range : Mango

Nature of damage : It attacks at flower bud burst stage and fruit set stage during January and May. Due to the attack of unopened flower buds, they fail to open and drop down. When the inflorescence stalk is attacked, the inflorescence becomes stunted and malformed.

Life history :

(i) Procysliphora tnangiferae (Felt) : The light orange coloured fly lays the eggs inside immature blossoms. The maggots that hatch out from the eggs feed on stalks of

stamens, anthers, ovary, etc. Only one maggot is found in each bud and it pupates inside the bud itself. The life-cycle from egg to adult occupies 12-14 days.

(ii) Dasineura amaramanjarae Grover : The adult flies insert the eggs into unopened flower buds. The maggots feed inside the buds and they fail to open and drop down. The maggots hibernate in the soil and thus carry-over of the pest to the next year is accomplished. When favourable conditions set in they pupate and emerge as adults.

(iii) Erosomyia indica Gr. : The maggots attack the inflorescence stalk, flower buds and small developing fruits. The adult fly is yellowish and lays the eggs on the infloresence peduncle or at the base of developing fruit. The maggots are yellowish and when full grown pupate in the soil.

Management strategies :

- Collection and disposal of infested panicles, leaves and twigs.
- Deep plughing of orchard in October-November to expose pupae and diapausing larvae to sun's heat which kills them.
- Monitoring of larval population on white paper in April/May and apply chlopyriphos (1.5%) dust based on population.
- Spray dimethoate (0.05%) at bud burst stage.

2.2. GUAVA

Guava (*Psidium guajava* L.) is one of the most important fruit crops in India. It covers around 3.3 per cent of the total area under fruit crops and contributes 3.3 per cent of the total fruit production in India. India is the major second largest producer of fruit in the world with 6383 million hectare area with annual production, of 74878 metric tons (Anonymous 2010-11). Bihar is the major guava producers than Uttar Pradesh, Karnataka, Madhya Pradesh, Gujarat, Andhra Pradesh and Maharashtra with annual production of 24.62 lakh tons, with area of 2,05,000 lakh ha.

The guava fruit is a berry with a large seedy core. The fruit may be smooth or ridge and waxy. Presently guava is being grown all over the sub-tropical and tropical world due to its high dietary value and good flavour. Guava fruit contains high amounts of Vitamins A, B1 (Thiamin), B2 (Riboflavin) and C. It is a rich source of vitamin C (Ascorbic acid). The vitamin C contents of guava fruit are four times higher than those of citrus. The leaves of guava have been used for curing diarrhea and dysentery. Guava is commercially picked when it starts turning from green to yellow so that it ripes one day later in the transit before marketing. Guava can be grown on a wide variety of soils from heavy clay to light sandy soil and with a range of pH 4.5 (acidic) to 8.5 (alkaline). It is tolerant to wet and saline conditions on good soil and with proper care the trees are highly productive. Guava produces high yield of good quality fruit in climates where there is a distinct winter season.

Guava is a very productive and highly profitable fruit crop. It is liked by fruit growers due to its wide adaptability and higher return per unit area. Raw guavas are eaten out-of-hand, but are preferred seeded and served sliced as dessert or in salads. More commonly, the fruit is cooked and cooking eliminates the strong odour. Bars of thick, rich guava paste and guava cheese are staple sweets, and guava jelly is almost universally marketed. Guava juice, made by boiling sliced, unseeded guavas and straining, is much used in Hawaii in punch and ice cream sodas. Green mature guavas can be utilized as a source of pectin, yielding somewhat more and higher quality pectin than ripe fruits.

Nutritive value of Guava

Nutrients	*Food Value Per 100 g of Edible Portion**	*Nutrients*	*Food Value Per 100 g of Edible Portion**
Calories	36-50	Phosphorus	17.8-30 mg
Moisture	77-86 g	Iron	0.30-0.70 mg
Crude Fiber	2.8-5.5 g	Carotene (Vitamin A)	200-400 I.U.
Protein	0.9-1.0 g	Thiamine	0.046 mg
Fat	0.1-0.5 g	Riboflavin	0.03-0.04 mg
Ash	0.43-0.7 g	Niacin	0.6-1.068 mg
Carbohydrates	9.5-10 g	Vitamin B3	40 I.U.
Calcium	9.1-17 mg	Vitamin G4	35 I.U.

*Analyses of whole ripe guavas.

2.2.1. Insect pests of Guava

More than 80 species of insects have been observed which, in one form or another, affect the quality and yield of guava; however few of them cause serious damage. However in India, the major pests are polyphagous pest like tea mosquito bug, fruit flies, fruit borers, bark eating caterpillars mealy bug and introduced pest namely spiraling white fly.

Major pests

Common name	*Scientific name*	*Family*	*Order*
Fruit fly	*Bactrocera* spp.	Tephritidae	Diptera
Fruit borer complex			
I. Pomegranate butterfly	*Deudorix (Virachola) isocrates* Fabricius	Lycaenidae	Lepidoptera
II. Castor capsule borer	*Congethes (Dichocrocis) punctiferalis* Guenee	Crambidae	Lepidoptera
Bark eating caterpillar	*Indarbela tetraonis* Moore	Metarbelidae	Lepidoptera

Other pests associated with guava

Common name	*Scientific name*	*Family*	*Order*
Tea mosquito bug	*Helopeltis antonii* Signoret	Miridae	Hemiptera
Mealy bug	*Ferrisia virgata* Cockerell, *Maconellicoccus hirsulus* Green	Pseudococcidae	Hemiptera
Coccids	*Hemaspidoproctus cinereus* Green	Margarodidae	Hemiptera
Guava aphid	*Aphis punicae* Passerini	Aphididae	Hemiptera
Fruit borer	*Rapala varuna* Hewitson	Lycaenidae	Lepidoptera

Fruit Fly, *Bactrocera* sp. (Tephritidae : Diptera)

Distribution : The fruit fly is the most destructive insect in the production of guava, particularly during rainy season. Guava fruits are attacked by a number of fruit flies including *Bactrocera cucurbitae, B. correcta, B. diversus* and *B. zonata.* These pests are widely distributed throughout the country. The main species, infesting guava in North India, have been identified as *B. Zonata, B. correcta* and *B. dorsalis.*

Host Range : Fruit flies are polyphagous pests, breeding on a number of fruits and vegetables and as such are of great economic importance.

Nature of damage : Ovipositional damage in the form of minute depressions may be seen from outside. Fruit soften at the site of infestation. The affected fruits rot and drop down prematurely. Ovipositional apertures lead to secondary infection by several pathogens.

Life history : Under, optimum conditions, a female can lay more than 3,000 eggs during her lifetime, but under field conditions from 1,200 to 1,500 eggs per female is considered to be the usual production. Eggs are laid on ripening fruits and the maggots feed on pulp of fruits. The mature larva emerges from the fruit, drops to the ground, and forms a tan to dark brown puparium. Pupation takes place in the soil. In North India, after the harvesting of mango, the pest shifts over to guava with the peak population in July-August, thereafter, population declines and larvae hibernate in winter.

Natural enemies : Parasitoids : *Opius compensates, Spalangia philippinensis,* and *Diachasmimorpha kraussi.*

Management strategies :

- To check the carryover of the pest, collect and destroy fallen and infested fruits along with fruit fly maggots.
- Tillage of tree basin also helps in checking the pest population as the pupae and hibernating larvae are destroyed by natural enemies.
- Conserve parasitoids such as *Opius compensates, Spalangia philippinensis, Diachasmimorpha kraussi* etc.
- Raking or ploughing the soil and application of *Metarrhizium anisopliae* @ 5 kg/ha to the soil underneath the tree canopy reduces fruit flies (Firake *et al.,* 2013).
- Use poison baits of Gur + Fruit juice 20% + Malathion 2% @ 40 baits/ha.

Fruit borer complex

Intensive surveys of guava growing regions of Uttar Pradesh revealed that fruit borer, *Deudorix isocrates* (Fab) has increased in incidence (2.5-22.5%) with crop loss range of 5.00 to 35.00 per cent. Common occurrence of another fruit borer, *Dichocrocis punctiferalis* (Guenee) in rainy season guava was also noticed. (Haseeb, 2006)

Pomegranate butterfly : [*Deudorix (Virachola) isocrates* Fab.] (Lycaenidae, Lepidoptera)

Distribution : The pomegranate butterfly, *D. isocrates* is a main pest of pomegranate, but recently infestation of this pest has been reported increasing in guava growing regions of Uttar Pradesh and other places in North India.

Host range : Aonla, apple, ber, citrus, guava, litchi, loquat, peach, mulberry, pear, sapota, tamarind.

Nature of damage : Larvae bore inside the developing fruits and feed on pulp and seeds just before the rind, exhibiting round bore holes on fruit. Infested fruits are also attacked by bacteria and fungi, which ultimately fall off and give an offensive smell. The larva bores the fruit and feed on the flesh and seeds, making the fruit hollow from inside. Infestation of this pest results in fruit loss. The entry and exit holes of the larvae also pave way for secondary infection by different pathogens.

Life history : The attack of this pest occurs in fruiting season both in rainy and winter season crops. The violet brown female butterfly lays shiny, white eggs singly on calyx of flowers and fruits. The larva is flattened, stout built, covered with short hair and about 2 cm long mature larvae may come out and pupate near the exit hole or may remain inside. Pupal period lasts for 7-34 days. Male is glossy, bluish violet, female brownish violet with an orange patch on forewings. Four generations are completed in a year.

Natural enemies : Parasitoids : *Trichogramma chilonis, Tetrastichus* spp., *Telenomus* spp., *Chelonus blackburni, Carcelia* spp., *Campoletis chlorideae, Bracon* spp.

Predators : *Chysoperla,* Rove beetles, Spiders, Coccinellids, Robber fly, Dragonfly, Reduviid bug, Praying mantis, Fire ants, Big eyed bugs, Pentatomid bug, Earwigs, Ground beetles, common mynah and King crow.

Castor capsule borer, [*Congethes (Dichocrocis) punctiferalis* Guenee] (Crambidae : Lepidoptera)

Distribution : South India (Major).

Host range : This borer is another polyphagous insect, larvae of which damage fruits of guava. It is primarily a pest of castor but also attacks guava, and other fruits, cardamom, jackfruit and forest trees and occasionally cause serious damage.

Nature of damage : Larvae of this moth mainly bore fruits but they may also bore buds and tender shoots. The affected fruits are generally deformed at the point of entry of larvae. Larvae faeces may be seen exuding out of the borer hole. Such fruits weaken, rot and drop down. The larvae feed on pulp and seeds of developing fruits resulting in premature dropping of fruits. The infestation of this insect takes place in rainy season guava.

Life history : Eggs laid on top leaf axils, inflorescence, tender part of plant and fruits, egg period 6-7 days. The mature larva is pinkish in colour, speckled with minute black spots, measuring about 1.5 cm in length. Larval period 12-16 days. The pupa is formed in pupal chamber made inside at the basal end of the fruit, pupal period 4-11 days. Medium sized bright orange-yellow color moth has numerous black dots on wings. Life cycle completed in 25-33 days.

Natural enemies : Same as mentioned in pomegranate butterfly.

Management strategies :

- Cultivation of pomegranate should be discouraged close to guava as this is the most preferred host of this pest.
- Detect early infestation by periodic monitoring. Install light trap @ 1/acre to monitor and mechanical collection of insects.
- Regular collection of infested fruits and their destruction to check the further spread and carryover the population.
- Conserve parasitoids and Predators.
- To prevent the infestation of pest, spray carbaryl (0.2%) or ethophenprox (0.05%) at the beginning of fruiting ason and before ripening of fruits. At least 15 days waiting period should be observed before beginning of harvest. The spray may be repeated at fortnightly intervals depending on severity of attack.

Bark-Eating Caterpillar [*Indarbela tetraonis* Moore and *I. quadrinotata* Walker] (Metarbelidae : Lepidoptera)

Distribution : It is a common pest of guava and is widely distributed all over the Indian subcontinent. Although, this pest is found in several parts of India like Bihar, Odisha, Haryana, Rajasthan, Madhya Pradesh, Maharashtra, Andhra Pradesh and Tamil Nadu, it is more common and destructive on guava trees especially in Punjab, Uttar Pradesh and South India.

Host range : Besides guava, the insect also infest mango, litchi, falsa, jamun, jack-fruit, pomegranate, ber and citrus plants.

Nature of Damage : The caterpillar remains hidden in the bore hole during day time. At night, it comes out and feed on the bark of the tree. For sometimes the larvae feed on the bark. The infestation of this pest may be identified by the presence of irregular tunnels and patches covered with silken web consisting of excreta and chewed up wood particles on the shoots, branches, stem and main trunk. Shelter holes may also be seen particularly at the joints of shoots and branches. A single larva inhabits a bore hole ; however, there may be 15 to 30 larvae in a single tree. The young shoots dry and die away giving sickly look to the plant.

Life History : Egg laying commences from April to June. Female lays eggs in cuts and crevices in the bark of the host tree in clusters of 15 to 25. In 8 to 10 days larvae hatches out of eggs. A fully grown larva is dirty brown in colour and measure 38 to 45 mm in length. The larval life lasts for 10 to 11 months. After that the larva pupates inside the galleries during March-April. Pupal period lasts for 15 to 25 days. Adult moths are white with pairs of small black dots on thorax, numerous small black spots and streaks on fore wings and few black spots on posterior edges of hind wings. There is a single generation in a year.

Management strategies :

- The frass and faecal matters along with ribbon like silken web should be scrapped from the bark and burn, so that caterpillars hidden under them are destroyed.
- Keep the orchard clean and healthy to prevent the infestation of this pest.
- Detect early infestation by periodically looking out for drying young shoots.
- Hot water may be inserted in the bore hole through syringe.
- After removing frass and faecal matter from tree bark, 0.1% emulsion of quinalphos or 0. 05% chlorpyriphos should be sprayed.
- Cotton wool soaked in carbondisulphide, chloroform or petrol should be inserted into the holes on tree bark. The opening of the hole may then be covered with mud.
- Kill the caterpillars mechanically by inserting the iron spike in shelter holes made by these borers at early stages of infestation.
- In case of severe infestation, remove the webs and insert the swab of cotton wool soaked in 0.05 per cent dichlorovos or inject water emulsion of 0.05 per cent monocrotophos or 0.05 per cent chlorpyriphos and plug the holes with mud.

Guava and Mango are two important fruit crops which are severely damaged by fruit flies. Most common species attacking, these two fruits are *Bactrocera correcta (Bezzi), B. dorsalis* (Hendal) and *B. zonnata* (Sounders), (Verghees and Sudhadevi, 1998 and Rajitha and Viraktamath, 2005). Female fruit flies have an ovipositor, similar to the (stings) of a wasp. With the ovipositor they puncture the skin of healthy fruit into which the eggs are laid (Marc *et al.*, 2010). Maggots come out from these eggs, pollute and destroy fruit by feeding on the pulp. Fruit fly is major limiting factor in production in rainy guava. Infestation of fruit fly ranged 20-46% with annual crop loss of 16-40% in U.P. (Haseeb, 2005). Maggots feed on the pulp of fruit make it unfit for consumption, therefore cause 25-50% loss every year (Qureshi *et al.*, 1976).

The intense application of synthetic chemical pesticides in different crops has created many complications such as pesticide resistance in insects, secondary pest outbreak and health hazards. Many alternatives are being proposed as eco- friendly and economically viable non, chemical method, biopesticide, biological control, botanical control, culture control or traditional pest management.

Therefore, in the present scenario, there is a need to reduce the environmental pollution and health hazards created by synthetic chemical pesticides.

❑❑❑

Chapter 3

Importance of Fruit Flies as Pest of Horticultural Crops

Dacinae fruit flies (Diptera : Tephritidae : Dacinae) are one of the key pests group of Asia and pacific, with the larval stage feeding on a wide range of fruits and vegetables. Direct fruit damage, fruit drop and loss of export through quarantine restrictions are all mechanisms by which fruit fly infestation causes economic loss. Adult fruit flies having high mobility and dispersive powers, high fecundity and in some species extreme polyphagy, Dacinae are well documented invaders and rank high on quarantine target list.

In many countries, the export of most commercial fruits is severely restricted by quarantine laws to prevent the spread of fruit fly species. Economic effects of pest species include not only direct loss of yield and increased management costs, but also the loss of export market and constructing and maintaining fruit treatment unit, having eradication facilities.

The major pest species belong to the genus *Bactrocera : B. cucurbitae, B. dorsalis* and *B. zonata,* while other species, such as *B. correcta, B. diversa* and *B. latifrons,* are still localized in their distribution. *B. versicolor* (Bezzi) (an earlier synonym of *B. dorsalis*), is very similar to *B. zonata,* but it differs from the latter in the presence of a black spot at the apex of the scutellum and in having a narrow (but complete) costal band. It is recorded from sapodilla (*Manikara zapota*) and guava. *B. nigrofemoralis* White and Tsuruta (known as *B. nigrotibialis* in the India literature) was reported from south India on *coffea canephoria* (Narayanan and Batra, 1960). It is known from South India and Sri Lanka and the only confirmed host is *Terminalia catappa* (Tsuruta and White, 2001). *B. (Tetradacus) minax* (Endertein) (also known as *Callantra minax* and *Polistomimetes minax* in the Indian literature), or Chinese citrus fly, is another fruit fly that needs monitoring. It is a pest of citrus and has already been recorded from northern Indian as causing heavy damage to tangerine (*Citrus reticulata*). *Carpomya vesuviana* Costa has extended its intensity of infestation on various parts of India, causing concern to the growers of ber (*Zizyphus mauritiana*) both in northern and southern India. The infestation is becoming so serious that sometimes not a single fruit is spared (Kapoor, 2005).

Besides fruit crops they are also destructive to many vegetables, oilseed crops and ornamental plants.

Female fruit flies lay eggs in fruits and ruin more than 400 different fruits and vegetables including mango, guava, citrus, melon, papaya, peach, passion fruit, plum apple and star fruit. They are strong flier and can fly upto two kilometers in search of food. Beside the direct damage of fruits, indirect loss is associated with quarantine restriction because infestation and sometimes mere presence of the flies in a particular country could also restrict the free trade and export of fresh horticulture produce to large lucrative markets. In India, a total loss of 2558 and 26,902 million rupees was estimated due to fruit flies with and without control measures, respectively (Stonehouse, 2001). Because of their infestation, India has been included in the list of those countries from where the import of fruits to developed countries has been banned. Important fruits flies damaging fruit crops in Uttar Pradesh include *Bactrocera cucurbitae, B. dorsalis, B. diversa, B. nigrotibialis, B. zonata* and *B. caudata* etc. Common host plant of *Bactrocera cucurbitae* are all cucurbitaceous crops. While *B. dorsalis* attack on mango, guava, peach, citrus, pear, ber; *B. zonata* infests guava, peach mango ber etc. (Kapoor, 1993). Fruit flies deposit their eggs in host fruits when they are physiologically ripe. On hatching, maggots bore their way to the interior and feed on the fruit pulp. Area fed by maggot is discoloured due to rotting of the fruit and the fruit drops prematurely. The methods of management of fruit flies are largely determined by their biological attributes. Only adults are exposed to control measures while eggs and maggots remain protected in the host tissues and most of insecticidal treatments are ineffective (Sharma *et al.,* 2011).

3.1. BIODIVERSITY

There are over one million known biological species out of which insects form almost three fourth of the total animal species biodiversity. The rate, at which new species have been discovered in the past, indicates that the actual insect species biodiversity is at least ten times more than what is presently known. Total world population of insects has been estimated at 10^{18} and even at most conservative estimate of 2.5 mg per insect, the total weight of insect will exceed that of Man by more than twelve times (Hocking, 1968).

3.2. DIVERSITY OF FRUIT FLIES

Fruit flies are the most diverse species on earth. Greatest diversity of fruit flies has been noticed in oriental region followed by afro tropical region. Diversity in pale arctic, Australasian, Neotropical and Nearctic region is almost similar. The least diverse fauna is found in Nearctic region. A great degree of regional endemism within the family is restricted to a single region. The total fruit fly genera show an average of 9.05 species/ genus.

Fruit flies comprise of about 200 genera. 183 genera of fruit fly are monotypic consisting of a single species while 6 genera have more than 100 species. These are *Anastrepha, Bactrocera, Campiglossa, Dacus, Tephritis* and *Trupanea.*

The genera which contribute the pest status of fruit fly are

1. *Bactrocera* (520 species)
2. *Dacus* (243 species)
3. *Anastrepha* (1980 species)
4. *Ceratitis* (78 species)
5. *Rhagaletis* (69 species)

Bactrocera spp. is pest of major importance in India. The common species reported on mango and other crops are *B. tryoni* (Frogatt), *B. zonata* (Saunders), *B. dorsalis* (Hendel), *B. neohumeralis* (Hardy), *B. jarvisi* (Tryon) and *B. frauenfeldi.* White and Elson-Harris (1992) regarded *B. dorsalis* on mango as a complex of 4 species.

There are about 325 species of fruit flies occurring in the Indian subcontinent of which 205 are from India alone (Kapoor, 2005).

3.3. INFLUENCE OF WEATHER PARAMETERS

Influence of weather parameters on fruit fly catches in guava, mango and aonla orchards was studied by Kumar (2018) and it was found that total fruit flies had highly significant positive correlation with minimum temperature. However, in guava there was significant positive correlation with II RH% and in second year significant positive correlation was registered with wind speed. In mango, total fruit fly population had highly significant positive correlation with maximum temperature and significant positive correlation with average rain fall during both the years. There was highly significant positive correlation with average rain fall and wind speed during 2013-14 and 2014-15, respectively. In aonla orchard during 2013-14 significant positive correlation with maximum temperature and highly significant positive correlation was registered with II RH%. During second year *i.e.* 2014-15 there was highly significant positive correlation with maximum temperature and wind speed.

3.4. ECOLOGY AND BEHAVIOUR

Fruit flies belong to Family Tephritidae, Order Diptera, of Class Insecta. The Tephritidae represents one of the largest families of Diptera which is most diversified and fascinating of Order Diptera. They are commonly referred to as "Fruit flies" due to their close association with fruits and vegetable. Sometime they are also known as "True fruit flies", as they lay eggs in mature rather than decomposed fruits. They are also known as "Peacock flies" due to their habit of strutting about, vibrating their wings and displaying their elaborate wings and body markings. Their ornate body and wing pattern serve as visual releasers in courtship and agonistic display.

The fruit flies infest stems, flowers, immature green as well as ripe fruits within their range of flight to ensure continual reproduction. The produce in the market or in transit is also attacked by flies. Their attack on the fruits not only reduces the yields but also affects the quality. The larvae may cause accidental gastro-intestinal myisis among human beings. Fruit flies of temperate areas have a comparatively low disperse capacity as compared to sub-tropical and tropical species. While temperate species are monophagous, sub-tropical and tropical species are polyphagous in nature and intend to increase their host range.

The insect population is regulated by many factors in nature such as abiotic factors, topographic factors and biotic factors. Among different factors that account for the natural balance of insects, the climatic or abiotic factors are perhaps the most effective. Temperature affects the insects directly and may have influenced a number of generations of insects in a given season. Extreme hot or cold weather will directly kill the insects. A wet climate is favourable to many insects during developmental period. Due to rain, the resultant wet, soft soil is essential for full grown maggots to enter into soil for pupation and also for the adults to emerge out of the pupae. In hard and dry soil, these life processes can not take place. Heavy rains, on the other hand, have adverse effect on soft bodied fruit flies. Strong winds interfere with host and mate finding flights and also the oviposition.

3.5. LIFE HISTORY

In general, the life history of many fruit flies is similar. Mating between the adult flies takes place at about dusk and last for about an hour or more. The pre oviposition period is 2 to 5 days but it may range from 10 to 15 days or longer in varying condition of climate and diet.

Egg

The female fly lays eggs in small clusters just underneath the skin, especially in the ripening of the fruit, 1-4 mm deep in the rind. Tactile bristles on the ovipositor assists the fly to discriminate between hard and soft surfaces for oviposition. As soon as a suitable spot on the fruit is located, it bends its abdomen to the right angle to the long axis of the body and thrusts the distal needle like structure of the ovipositor and the proximal tube deeper and deeper until it enters completely into the tissue and the eggs are then deposited. After oviposition, the ovipositor is withdrawn leaving a puncture behind. The complete act of ovipostion takes some time about 5 minutes. White banana shaped eggs are deposited in clusters. Each cluster may contain 200 to 1,500 eggs depending upon the texture and quantity of the pulp available in each kind of the fruit. On an average about 50 eggs are laid but a single female may lay about 150 to 200 eggs in its life time in a month. The incubation period of eggs is 2 to 3 days during March and April and even 24 to 36 hours in summer months. It may be longer up to 10 days in winter.

Maggot

The young larva is white, translucent and measures 1.5 mm × 0.3 mm. The full grown larva is 8 to 9 mm long and 1.5 mm broad. The larva turns yellow at a later stage due to accumulation of reserve fats and becomes opaque. The head of the cephalic segment is pointed anteriorly. On the dorsal surface of the head two anteenal protuberances and a pair of maxillary palps are found. A cephalopharyngeal skeleton is found with sclerotised mouth hooks. The last abdominal segment consists of red stigmata forming the posterior spiracles. A pair of interior spiracles is however present on the first thoracic segment. Normally there may be 2 to 12 maggots in a fruit, but in case of shortage of host, one fruit may have ten times more number of maggots. The larval period lasts for 6 to 29 days depending upon the period of the environmental conditions. There are three instars and the fully developed learva (third instar) has a habit of jumping a short distance to find a suitable place for pupation. The pre-pupal period ranges from 18 hours to 2 days.

Pupa

Pupa is cylindrical in shape, 4–5 mm long and 2 mm broad. The colour varies from dull deep reddish yellow to ochraceous. It is exarate, encapsulated within a puparium. Puparium may be smooth or wrinkled with distinct 11 segments, the last segment being a little more prominent. The interior spiracles are much darker but similar to those of larva. The posterior spiracles are situated below the mid horizontal lines and are yellowish in colour. Each spiracle has 3 small but relatively broad opening. The pupal period lasts from 6 to 44 days. Generally, pupation takes place 7 to 17 cm below the soil surface. Humidity plays an important role in the development of pupa. Humidity above 85% is required for the development of adult inside the pupa.

Adult

A well developed adult is stout and a little bigger than the ordinary house fly. It measures 14 mm across the wings and 7 mm in length. The male fly is slightly smaller in size than the female. Female has a tapering abdomen which ends in a pointed ovipositor. The fly is generally brown or dark brown with hyaline wings and brown to yellow legs. Wings are peculiarly marked with patterns and streak along coastal band is found in many adults.

3.6. BACTERIAL ASSOCIATION

The most interesting aspect of nutrition and survival of larvae and adults is their association with symbiotic bacteria. The bacteria multiply in elementary canal. A pseudomonas type bacteria was identified from eggs, larvae and adults of *B. cucurbitae.* The three most important bacterial spp. in *B.tryoni* were *Erwinia herbicola, Enterobactor cloacae* and *Klebsiella oxytoca*. The fly regurgitates fluid containing bacteria on to the fruit surface and oviposits through it. The bacteria might help in the break down and digestion of fruit tissue.

3.7. LARVAL BEHAVIOUR

During the development, the larva tunnels in the fruit, macerates the tissues and ingests the broken down tissues and associated bacteria. When full fed, larvae of most species leave the fruit and burrow many centimeters deep into the soil where they pupate. Mature larvae of all species have the ability to hop for scattering them and it also appears to be defense mechanism against ground dwelling insect predators particularly ants. The larval exit usually occurs at dawn and is usually controlled by light-temperature cycles.

❑❑❑

Chapter 4

Adult Habits

4.1. MOVEMENTS

Polyphagous mutivoltine, tropical and sub-tropical species *i.e. B. cucurbitae, B. dorsalis* and *B. zonata* are strong flyers and are highly mobile. They engage in extensive dispersive movements during host seeking and mating. Mature adults fly from site of dwindling hosts to new host. Maximum distance recorded for marked individuals was 200 km for *B. cucurbitae*, 65 km for *B. dorsalis* and 90 km for *B. tryoni*. When the hosts are in plenty in an area, mature flies tend to restrict their movements to foraging flights in search of food, water and ovipositional sites. The overall movements are non dispersive. Flies may show a daily pattern of movements between hosts and surrounding vegetation in *B. cucurbitae*, which enters cucurbit plots at fairly specific time of day to oviposit and then return to the nearby vegetation to feed and rest (Bateman, 1972).

4.2. FEEDING AND HOST SELECTION

Adult Decines have been observed feeding hosts and non hosts, honeydew, plant exudates, extra floral nectarines, pollens, fruit juice, ripe fruits, microorganisms and bird droppings. Peak feeding normally occurs in the morning, but some feeding may take place at other times. Adult flies use visual and olfactory stimuli to locate host. The responses of fruit flies to visual stimuli are dependent on colour, shape and size of the stimulus.

4.3. SEXUAL BEHAVIOUR

Sexually active males of *B. dorsalis* and *B. cucurbitae* produce a high pitched buzzing sound by rapidly drawing their wings across their stridulatory organs and simultaneously release pheromones from their rectal glands. The majority of dacine fruit flies including *B. dorsalis* and *B. cucurbitae* mate at dusk under low light intensity. Some start mating in the late afternoon at a somewhat higher light intensity and few even mate during the day at high light intensity.

4.4. ADULT MATURITY AND MUTING SYSTEM

Dacine fruit flies are anautogenous, *i.e.* they immerge from puparia as sexually immature adults that need to forage for resources to facilitate survival and reproduction. Key resources include moisture or metabolism, sugars to get strength for flight, protein to attain sexual maturity and lipids for egg production. Sugar sources include honey dew and other plants exudates. Protein is obtained from phylloplane bacteria and bird feces and moisture is derived from dew and rain. Female flies are considered monandrous as regards to the frequency of mating, while male flies are polygynous. Mating occurs within a defined special arena where resources for adult flies are absent. In such arenas, sexual selection by the females' preference for males is done for those males that have fed on plant derived phenylpropanoids. These plant chemicals when injected are integrated into the male fly's sex pheromone which makes males more attracted to females. The males that have had prior exposure to methyl eugenol acquire more mates than males that have not.

4.5. RESTING AND SHELTERING

Adults normally rest on the under sides of leaves of both host and non hosts at night and in day time during periods of inactivity. When day temperature rises above 35°C, adults of *B. zonata* have been observed to leave. Large numbers of both sexes may aggregate in these roots during the day to feed but return before nightfall.

4.6. MATING BEHAVIOUR

These flies are anautogenous which means that they emerge from puparia as sexually immature adult that forage for resources to facilitate survival and reproduction. The most important resources include moisture for metabolism, sugar to fuel flight, protein to attain sexual maturity in conjunction with lipids for egg production. Sugar sources include honey dew and other plant exudates. Protein is derived from phylloplane bacteria and bird feces and moisture from dew and rain. Adult flies forage for these resources. Certain plant derived phenyl propanoids (*e.g.* Methyl eugenol and raspberry ketone) are also required which play an important role in the mating behaviour of these flies. Female flies are considered monandrous while males are polygynous. Those males which have fed on plant – derived phenylpropanoids are preferred by females. The ingested plant chemicals are integrated into male fly's sex pheromones which makes them more attractive to females. Males and females of *B. dorsalis* aggregate on the host which had prior exposure to methyl engenol. Such males acquire frequent mating as compared to males that had not fed on such plant derived phenylpropanoids.

4.7. INTRA SPECIFIC AND INTER SPECIFIC COMPETITION

In *B. oleae* only one individual normally completes development, irrespective of the quantum of eggs laid per fruit by the fruit fly. Oviposition deterring pheromone, which checks the over exploitation of the larval food have only been found in *Anastrepha* spp. and *C. capitata* but not in *Bactrocera* spp. In polyphagous species, scramble competition may operate over a wider range of larval densities and it operates between larvae of single age class, while the older larvae have a highly detrimental effect on younger larvae in the same fruit. (Fitt, 1987). In guava, there was no difference in the larval counts/fly emerged per egg puncture even when there were 9 egg punctures per fruit.

Interspecific competition has been observed in native and introduced species or in two introduced species occurring on same hosts. In such cases, certain species *e.g. B. dorsalis* seems to be superior competitor when in contact with *C. capitata*. In *B. dorsalis* a type Cleptoparasitism (*i.e.* the use of egg punctures of *Ceratitis* for oviposition) and inhibition of the larval development of the competitor (Christenson and Foote, 1960) are special adaptations of the polyphagous spp to eliminate the older fruit fly specimens in a given host.

4.8. POPULATION FLUCTUATIONS

The abundance of larval hosts is one of the major factors regulating dacine populations. In the poyphagous species, populations are usually at their lowest level at the end of the winter owing to the slow down or cessation of breeding and it may take several generations for them to built up to peak level. Minimum population levels generally occur just prior to infestation of new crop.

Heavy rains and long period of rainy days is also deterimental of fly populations. High temperature during summer causes high mortality of eggs and young larvae in fruit and population in soil. Natural enemies and intraspecific and interspecific competition also regulates the dacine population in few cases.

4.9. ABIOTIC FACTORS

Studies on abiotic factors conducted recently on fruit fly catches in guava, mango and aonla orchards, by Kumar (2018) revealed that fruit flies had highly significant positive correlation with minimum temperature. However, in guava there was significant positive correlation with II RH% and in second year significant positive correlation was registered with wind speed. In mango, total fruit fly population had highly significant positive correlation with maximum temperature and significant positive correlation with average rain fall during both the years. There was highly significant positive correlation with average rain fall and wind speed during 2013-14 and 2014-15, respectively. In aonla orchard during 2013-14 significant positive correlation with maximum temperature and highly significant positive correlation with II RH% was recorded.

4.10. BIOTIC FACTORS

Six parasites have been recorded on *B. cucurbitae* in India and among these *Opius fletcheri* is predominant. But all these parasites were unable to check the fly. However only 20% parasitism is reported by *O. fletcheri.* Five braconids have been also recorded on *B. dorsalis.*

Ecological factors play an important role in the population dynamics of different fruit fly species. Ecological factors like temperature, RH are positively correlated with population density of cucurbit fly, while the incidence of *C. vesuviana* is negatively correlated. Cloudy weather favours the population built-up of various fruit flies. Extreme hot and cold seasons adversely affect fruit fly population in nature.

❑❑❑

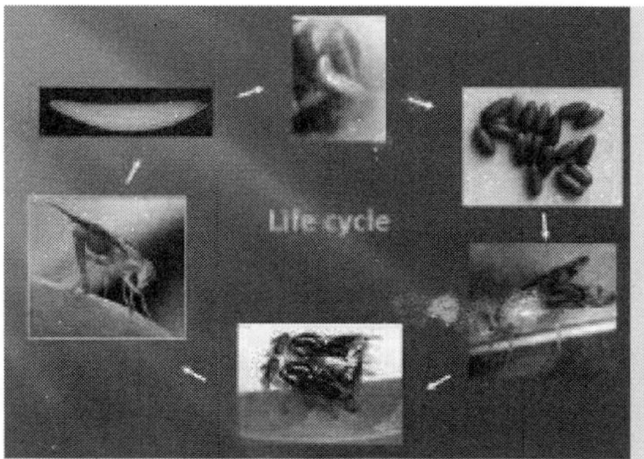

5

Chapter

Classification

Fruit flies belong to the family tephritidae, which is one of the largest, most diversified and fascinating acalypterate family of this order. These are commonly called as fruit flies due to their close association with fruits and vegetables. They are also known as peacock flies because of their habit of strutting about and vibrating spotted and striped wings. Of the 4500 known species of fruit fly worldwide, nearly 200 are considered as pests but about 70 species are regarded as agriculturally important throughout the world (Clark *et al.,* 2005). David and Ramani (2011) reported that 325 species of fruit flies are known to occur in Indian sub-continent of which 243 in 79 genera are from India alone. The tribe dacini with genus *Bactrocera* is of importance in India and from the economic point of view, oriental fruit fly or mango fruit fly *Bactrocera dorsalis* (Hendel), guava fruit fly *Bactrocera correcta* (Bazzi) and peach fruit fly *Bactrocera zonata* (Saunders) are very important pest of fruit crops and are recognized worldwide as the most important threat to horticulture. *Bactrocera dorsalis* is considered to be among the five most damaging and aggressive fruit flies in the world (Lablank and Putoa, 2000).

Genus *Bactrocera/Dacus* belongs to :

Order : Diptera

Sub-order : Brachycera

Infra-order : Muscomorpha (= Cyclorrhapha)

Section : Shizophora
Super-family : Tephritoidea
Family : Tephritidae

Order Diptera : Diptera has the following characters :

A. Single pair of wings, the hind pair modified into haltares.
B. Mouth parts suctorial forming a proboscis, labium distally expanded into a pair of fleshy lobes.
C. Larvae dipteroid cruciform, cylindrical, segmented, apodous; mouth parts inconspicuous.

Family Tephritidal = (Trypetidae) (Fruit flies, pictures-winged flies) : Family tephritidae has the following characters :

A. Wing as a rule conspicuously marbled.
B. Small to medium sized flies 4.5-10 mm long, slightly hairy and bristled. Usually yellow amber, orange, black and combination of these colours.
C. Ovipositor horny, flattend and prominent, three segmented, larvae muscoid shortened when mature.

Family Tephritidae is further classified in two sub families of economic importance :

(a) **Sub family** : **DACINAE**
  (i) **Tribe** : DACINI
    Genus : (A) BACTROCERA
    Subgenus : *Bactrocera, Hemigymnodacus, Javadacus and Zeugodacus*
    Genus : (B) DACUS
    Subgenus : *Callantra*
  (ii) **Tribe** : GASTROZONINI
(b) **Subfamily** : **TEPHRITINAE**
  (iii) **Tribe** : PLIOMELAENINI
    Genus : *Pliomelaena* Bezzi
  (iv) **Tribe** : TEPHRITINI
    Genus : *Dioxyna* Frey

It comprises of a large group approximately 5000 spp of moderate sized flies. They are completely worldwide in distribution but more abundantly found in tropics and warmer regions. But, they should not be confused with Drosophilidae.

5.1. DISTINCTIVE CHARACTERS

- Medium sized flies, upto 35 mm in length.
- Costa with humeral and costal breaks.
- Anatomical characters include two small breaks in the costal vein of the anterior margin of the wing above the second costal cell, sometimes allowed to R1.
- Subcosta bent sharply forward at right angel before apex. After the bend it is weakened or folded.
- Vein R1 densely setulose dorsally.
- Cell cup with a posterior apical extension.
- The body of the fly is often brightly coloured, ornate but non metallic.
- The colour pattern is an important taxonomic character.
- Wings are often spotted (maculate) or marbled.
- The flies generally walk slowly and wave its wings back and forth.
- Female has a long short, extendible, horny ovipositor (oviscape) used to lay eggs within host fruit or other plant parts.
- Female has a characteristic oviposition deterrent or host marking pheromone.
- Males often have a comb (pecten) of bristles on the edge of the third abdominal tergite and a pair of large, oval shining spots on the fifth tergite.
- Adult flies feed on sugary solution, ripe fruits, honey dew and without food can live for 5-6 months.
- This is the reason probably they are important as crop pests.
(Agrawal and Mishra, 2010)

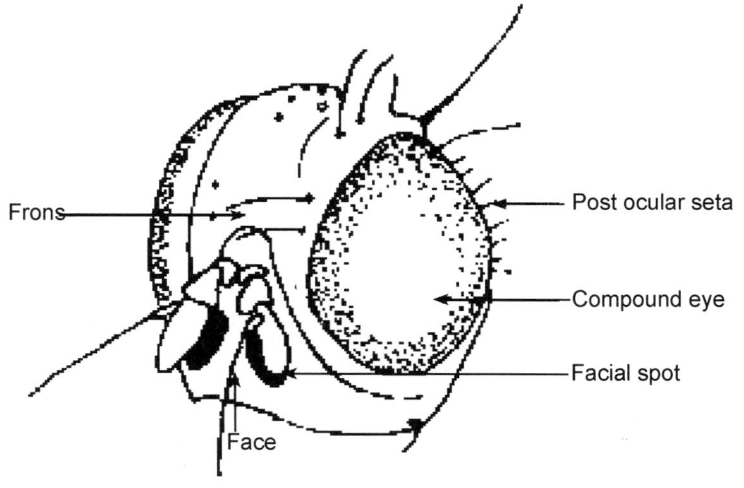

FIG. 1. Head

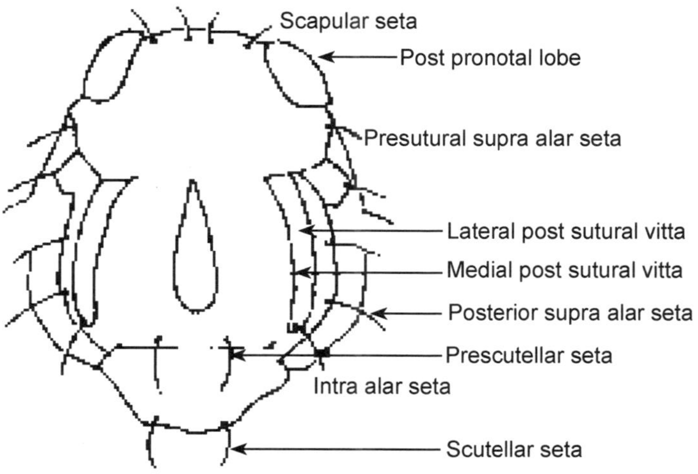

FIG. 2. Thorax

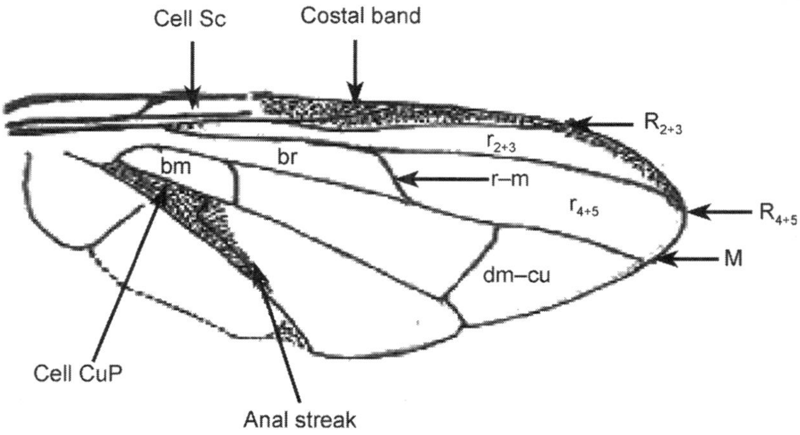

FIG. 3. Wing

ABDOMEN

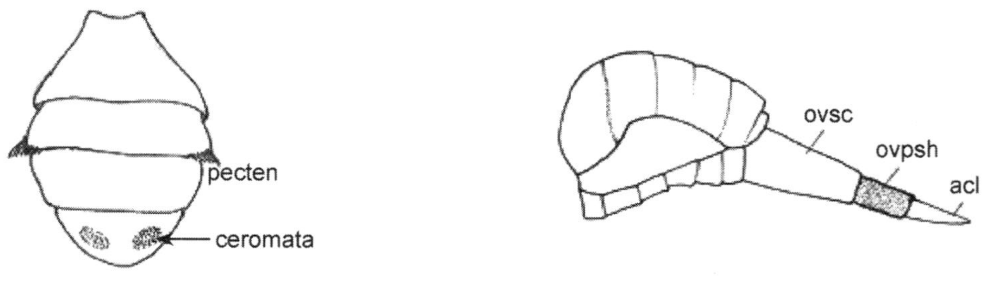

FIG. 4. Male **FIG. 5. Female**

5.2. PEST STATUS

The larvae are mostly phytophagous with three larval instars. Third instar larva is useful in identification. According to their feeding behaviours they can be grouped in to the following types :

1. Inside flower heads of compositae : *Tephritis*
2. Inside fleshy fruits : *Ceratitis, Bactrocera* etc.
3. Leaf miners : *Philophylla* etc.
4. Stem miners/bores : *Paroxyna misella*
5. Gall formers : *Procecidochares, Urophora* etc.

Tephritid larvae cause serious damage to agricultural crops that feed inside developing fruits etc. Infested fruits often fall prematurely. The female fly punctures the outer tissue of plant or fruit and deposit eggs with its ovipositor under the skin. The larvae develop inside the fruit which may or may not fall prematurely. Eventually the fully grown maggots leave the fruit and pupate in the soil. The oriental fruit fly is one of the most serious pests of mango in the country which has created problem in the export of fresh fruits. *Bactrocera dorsalis, B. zonata* and *B. correcta* are the most common fruit flies which cause serious damage to mature mango fruit.

Most of the crop pest species are polyophagous and may attack a wide range of cultivated fruits, a number of fruit flies are oligophagous while a few have been recorded from a single host. Many plant families have been chosen as hosts by fruit flies, but most important are Compositae, Rosaceae, Cucurbitaceae and Rutaceae. It is possible that locally infested fruits may have several different species of the fruit fly. Usually each individual fruit is only attacked by one species of fruit fly but the crop as a whole may be infested by several species.

In India mango and cucurbits both have been recorded six species of Tephritids infesting fruits. 4 species on mango have been reported in Karnataka (Babu and Viraktamath, 2003; Ranjitha and Viraktamath, 2006) and in Uttar Pradesh (Agrawal and Deepa M, 2008) such as *Bactrocera dorsalis, B. correcta, B. zonata* and *B. cucurbitae.*

Some of the fruit fly species are restricted to certain geographical region and due to their status as important fruit pest; they have become the subject of international quarantine legislation. Since these flies are so easily transported in infested fruits and other ways, it needs continuous process of monitoring fruit flies to limit their geographical spread. In 1974 *Dacus (Bactrocera) dorsalis* became established in California and in 1975 in Northern Australia. Mediterranean fruit fly *Ceratits capitata* has been accidentally introduced in Maxico.

5.3. TEPHRITIDS AS BIOCONTROL AGENTS OF WEEDS

Most of the world's noxious weeds belong to the family compositae. In recent years a number of host specific Tephritids have been utilized in biocontrol programmes in Australia and other parts of the world. Some success in this field has been achieved. Crofton weed *Ageratina adenophora* has been substantially controlled by introducing Tephritinae *Procecidochares utilis* (Orgion : Mexico) in 1963 in the Nilgiri area of Darjeeling and Kalimpong areas. The fly had been established well and spread naturally and rapidly. It failed due to heavy rainfall and parasitism due to indigenous natural enemies. But it is possible that flies of this family will be used more widely in the control of weeds.

5.4. IDENTIFICATION

Earlier classifications of fruit flies had been on trivial characters. Due to gradual increase of the fruit fly taxa based on such characters, a situation had developed that it became difficult to assign correct names to many of the fruit fly specimens. With the passage of time the fruit fly workers started realizing this complex problem and made noticeable improvement not only in the usage of reliable characters for the recent taxa but also revised the earlier descriptions. The present day taxonomic characters thus include important morphological features (including genitalia). An attempt has thus been made to identify the fruit flies which have been found dominant in India, infesting cucurbits and fruits.

5.5. SYSTEMATIC POSITION OF GENUS *BACTROCERA (DACUS)*

Genus *Bactrocera* (*Dacus*) belongs to :

1. Order : Diptera

2. Group : Cyclorrhapha

3. Sub-order : Schizophora

4. Series : Acalyptratae

5. Family : Tephritidae (Trypateidae) and has the following characters :

(A) Wing as a rule conspicuously marbled.

(B) Small to medium sized flies, slightly hairy and bristly usually yellow amber, brown, orange, black and combination of these colours.

(C) Ovipositor horny, flattened and prominent, three segmented, muscoid, shortened when mature.

6. Sub-family : Dacinae

According to food habits of the larvae (Essig) has given the four categories of the family Tephritidae :

(A) Living in fruits preferably of 'fleshy type'.

(B) Living in flower heads of compositae.

(C) Leaf or stem minors.

(D) Gall formers on the various parts of the plants.

7. Genus (*Dacus*) : *Bactrocera* belongs to the first group.

❑❑❑

6

Chapter

General Morphology

6.1. HEAD : (FIG. 6 & 7)

General shape of the head is sometimes important : broad or narrow, windend below or not, with the face retracting inferiorly or not.

(i) Frons : Also, called front, it is the upper anterior part of the head, between the eyes from the vertex to the base of antennae; the distance between the black ocellar triangle on the vertex and the base of antennae is the length of the frons and the width is between the eyes or sometimes, if necessary, between the antennae; its pubescence is sometimes taken as an important generic as well as specific character.

(ii) Antero-medial hump : It is a small hump or protuberance almost in the centre of the frons.

(iii) Superior or upper fronto–orbital bristles : One or two pairs (always one in Dacinae) placed on the most dorsal part of frons; which are bent backwards, and are rarely reduced or absent.

(iv) Inferior or lower fronto-orbital bristles : One to four pairs, rarely six to eight pairs, on frons below superior orbitals; they are bent forward or outwards; and are rarely reduced or absent.

(v) Frontalia (Frontavitta) : It is the middle portion of the frons; often baring slight, inconspicuous to dense and coarse pubescence; and is sometimes quite bare.

(vi) Lunule or frontal lunule : It is a small raised area interiorly just before the antennae; and its shape is sometimes useful.

(vii) Vertex : It is the top most part of the head between the eyes, frons and occiput.

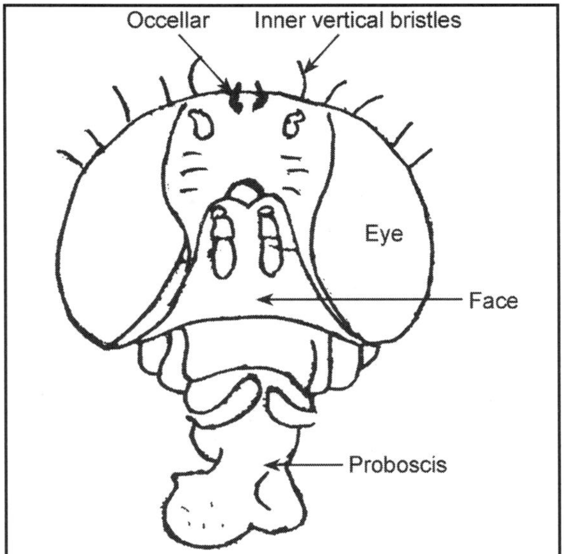

FIG. 6. Head (Dorsal view)

(viii) Ocellar triangle : In the middle of the posterior end of frons is a more or less clearly defined triangle, with its base on the vertex; this bears a rounded somewhat raised area in the centre, usually black in colour called the ocellar dot; this dot nears three ocelli and the proclinate ocellar bristles. The presence or absence of these bristles is sometimes an important generic as well as specific character.

(ix) Face : It is an inverted v-shaped area on the anterior surface between the antenna and epistoma. It often contains two dark oval spots or elongated bands on each side below the eyes or sometimes only a broad transverse stripe or the face sometimes is entirely black, yellow etc. The presence or absence of spots, stripes, etc., is a valuable specific character.

(x) Epistoma : It is the ventral margin of the head; or is also called lower facial margin.

(xi) Gena : It is the subocular region between the eye and oral margins on either side; and it often bears a bristle called the genal bristle.

(xii) Subocular spot : It's a dark spot on each gena adjacent to the eye margin.

(xiii) Inner vertical bristle : A pair, long, on vertex.

(xiv) Outer vertical bristle : A pair, short, on vertex.

(xv) Post verticals : A pair, short and weak, on vertex, behind ocellars (so also called post-ocellars), is parallel or diverging, and is rarely absent.

(xvi) Occiput : It is the whole posterior area of the head below and behind the eyes.

(xvii) Occipital row : They are a row of black, thin, pointed bristles, or whitish, thick, truncated bristles on the occiput along the posterior margin of each eye; sometimes quite a useful taxonomic character at sub-familial and generic levels.

(xviii) Compound eyes : They occupy the greater part of the sides of the head; may be oval sloping as a rule a little back-wards, or may be large in oval heads and smaller in angular heads; their shape and size are sometimes important generic characters.

(xix) Antennae : Three segmented, the first joint is short, disc like and inconspicuous, is rarely long and prominent; the second joint is short and cap like, covering the proximal end of the third joint, occasionally may be elongated; the third joint is longer, longer than the first or the second joint or both combined. Sometimes more elongated reaching the epistome or beyond. The ratio of such combined length of the head is an important generic and is sometimes a specific generic character. The third joint bears an elongate bristle, the arista, on the dorsal side of the proximal region; it may be bare, pubescent, and two to three segmented. The ratio of its length to the antennal joints and it being bare or with hairy covering are useful generic and specific characters.

(xx) Proboscis : It is mostly a soft structure; its shape and size is seldom taken as a taxonomic character in the separation of genera and species.

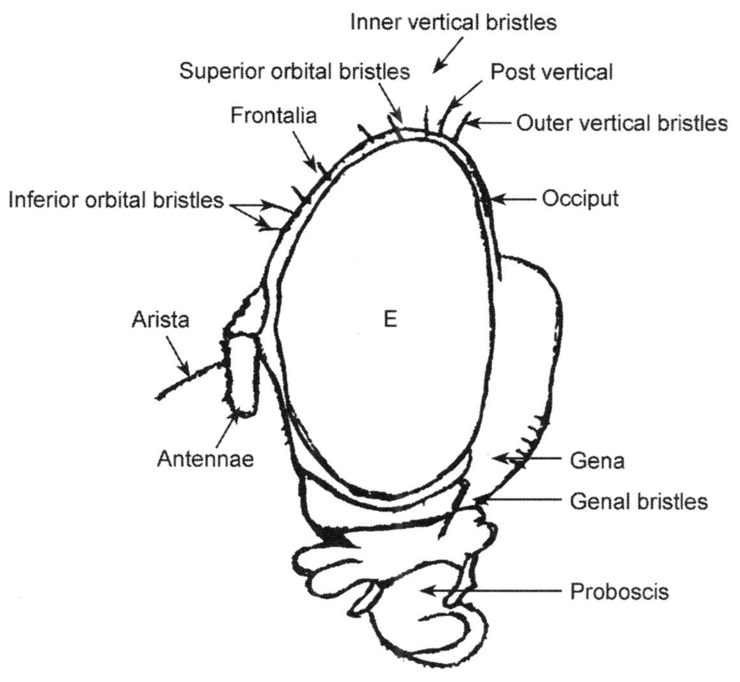

FIG. 7. Head (Lateral view)

6.2. THORAX : (FIG. 8 & 9)

(i) Humeral calli : They are rectangular or rounded protuberances on the anterolateral corners on the dorsal surface of mesothorax. Each callus bears a single humeral bristle which is sometimes absent. Its colour and chaetotaxy sometimes are important taxonomic characters.

(ii) Scapular bristles : They are two pairs, small, and are on the anterior margin of mesonotum; sometimes less distinct but never absent; one pair is median or achrotichal and one pair is lateral or dorso-central.

(iii) Propleuron : It is one of the lateral parts of the prothorax; it bears one or more pro-pleural bristles or hairs and thus sometimes is taken as an important generic and specific character.

(iv) Notopleural calli : They are also called post-humeral calli; each is raised, and has somewhat triangular protuberances on the notopleurae between the humeral calli and the wing base. It bears a pair of bristles on each notopleural area – the anterior bristles is at about midway between humeral and notopleural calli while the posterior one is on the apex of the notopleural callus.

(v) Pre-sutural bristle : It is a long bristle on either side of mesonotum just behind the anterior notopleural bristle; and is sometime, absent.

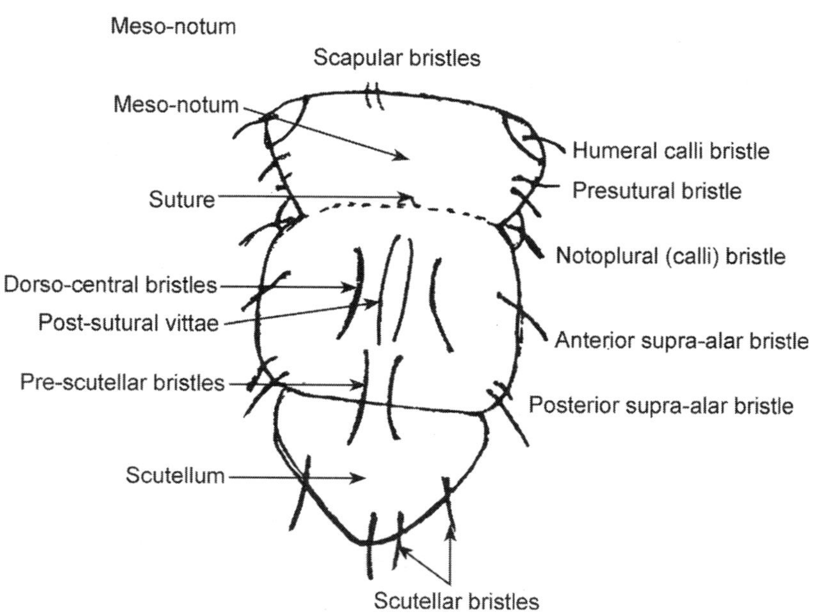

FIG. 8. Thorax (Dorsal view)

(vi) Hypopleural calli : Also called pleuroterga or pleurotergal calli; they are two rounded raised protuberances on each side of the thorax posteroventral to the wing base.

(vii) Mesopleuron : It is the lateral surface of the mesothorax. It bears one or two bristles on the posterior border below each notopleural callus, sometimes accompanied with some bristly hairs; divided longitudinally into on upper episternum and lower epimeron which is in turn transversely divided into anepisternum and anepimeron and katepisternum and katepimeron.

(viii) Mesopleural strip : It is a yellow band on each lateral surface of the mesopleuron, the anterior margin of the pteropleuron and it usually continues into the upper margin of the sternopleuron. The shape and size of the stripes, especially their width in relation to the humeral calli, are important taxonomic characters at the species level.

(ix) Pteropleuron : It is the posterior lateral plate below the base of the wing behind the mesopleural suture. It bears one bristle, under the root of the wing, usually strong but sometimes weak.

(x) Sternopleuron : It is the region below the mesopleuron and pteropleuron and above the fore and hind coxae on each side of the thorax. It bears one strong bristle below sternopleural suture, and is very rarely absent. Its presence or absence is a useful generic character.

(xi) Post-scutum : It is the largest area of the mesonotum behind the suture.

(xii) Anterior Supra-alar bristles : These represent a pair; they are small and are just behind the suture directly above the base of each wing on the lateral margin of mesonotum; they are sometimes used as the base line of the dorsocentrals; and rarely absent.

(xiii) Posterior Supra-alar bristles : These are two pairs, one above the root of the wing and one on the postalar callus (also called postalar bristles); in Dacinae one bristles is dorsal to the other and in line with the lateral postsutural vita; and is always present.

(xiv) Pre-scutellar bristles : These represent a pair, and are on the posterior margin of the mesonotum just anterior to the scutellum; these are strong, weak or may be absent.

(xv) Post sutural vittae : These are three in all, two are lateral, and one is median; usually in Dacini, one is on each lateral margin and a single median is running along the mid-line of the mesonotum; these are mostly yellow in colour and the stripes or bands vary in shape and size; laterals may begin at or anterior to the mesonotal suture and end before, at or posterior to the upper posterior Supra-alar bristle; often a lateral yellow band present connecting humeral in nototpleural calli in Dacinae.

(xvi) Scutellum : It is a triangular section cut off by an impressed line from the posterior margin of the mesonotum. It bears two to four or six rarely 10 pairs of bristles; when only two pairs – one pair is apical and one pair is on the basal side where the scutellum meets the posterior margin of mesonotum. Its shape, size colouration and chaetotaxy are important taxonomi characters at generic and species level.

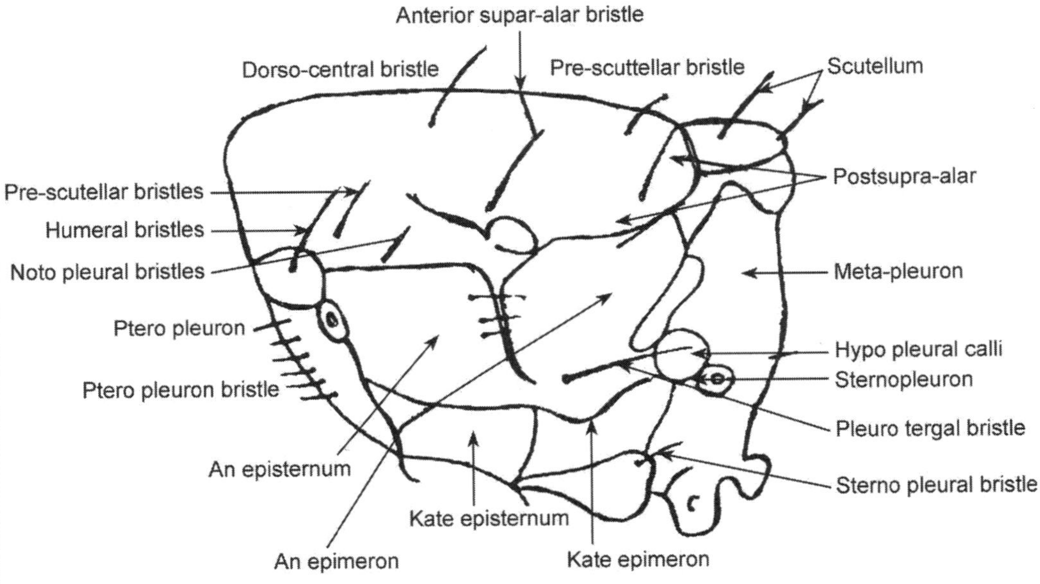

FIG. 9. Thorax (Lateral view)

6.2.1. Legs

The shape, size and chaetotaxy of major leg segments (coxa, femur, tibia and tarsus) offer useful generic characters. Sometimes colour is also a useful specific and generic characters. The number of apical spurs on the mid-tibiae also offers useful generic character.

6.2.2. Wings : (Fig. 10)

The shape, size and colour pattern and venation of the wings are important taxonomic characters at all levels in fruit fly taxonomy. The wing venation is reduced due to the fusion of some veins. Thus, the main veins are Costa, Subcosta, three Radials (R_1, R_{2+3} and R_{4+5}), two Media (M_{1+2} and M_{3+4}) and one representing the fusion of cubital and Anal veins (Cu_1 + IA) bearing a dense mass of microtrichia in males but never in females. There are less number of cross-veins too, of which two are important. One r-m in between R_{4+5} and M_{1+2} or on discal cell. Its position on the discal or discoidal cell is a very useful generic and specific character. Another cross-vein is in between M_{1+2} and M_{3+4} forming the apical part of the cell. The ratio of its length with r-m is sometimes a useful character. The presence or absence of spurs or setae on longitudinal veins (R_{2+3}

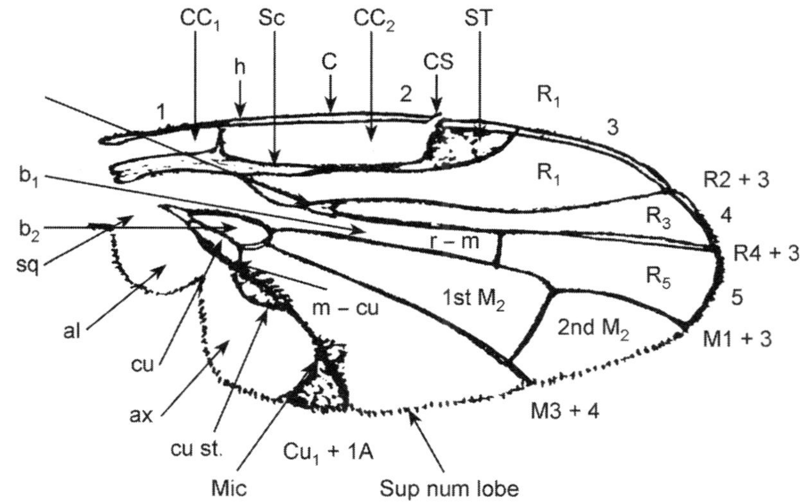

FIG. 10. Wing

and R_{4+5}) and pattern of the veins sometimes offer important generic and specific characters.

The wing veins offer the most valuable family characters like the presence of two breaks in the costal vain on the anterior margin of the wing above the second costal cell; and the short, fine subcostal vein in the costal cell; and the short, fine subcostal vein in the costal cell but ending much before the wing margin at right angle; and the node at the junction of R_{2+3} and R_{4+5}.

6.2.3. Wing cells

The various cell and the supernumerary lobe are also very important taxonomic characters. The usual cells in a wing are two costal cell (1C, 2C), cell R_1, R_3, $1R_5$, $1M_2$, $2M_2$, M_4 and 'a' (anal cell). The shape and size of the discal cell ($1M_2$), supernumerary lobe and that of the anal cell are important generic and specific characters. Sometimes the presence or absemce of a mass of small microtrichia withing the costal cells is a specific character, two, occasionally three costal spines present at the end of the subcosta, the upper one may be short and inconspicuous. The supernumerary lobe is also an important charcter in the speparation of various subgenera of the genus *Bactrocera*. It is reduced in size in the males of some and females of all Dacinae species.

6.3. ABDOMEN (FIG. 11 & 12)

It may be rounded, oval, elongated, club-shaped or even petiolate; the terga are free. In African species, these are fused into a hemispherical syntergum. In the male, there are five well developed segments; segments VI to VII are absent, segment VIII may be absent or become a part of male genitalia and segment I exists as a reduced genital segment. Segment IX and sternite V of male are important characters at generic level in Dacinae. In the female abdomen, there are I to V well developed segments, segment VI is present while segments VII to IX are modified to form the ovipositor. In the males of most species, especially Dacinae, is present a pectin or comb of cilia on the postero-lateral corners of tergite III, and also a pair of large oval shining spots on tergite V. The shape of abdomen together with fusion or non-fusion of its tergites, presence or absence of pectin on III tergite, ratio in size of V and VI segments of female, shape of abdominal sternites I and V in male and colour pattern are also taxonomic characters.

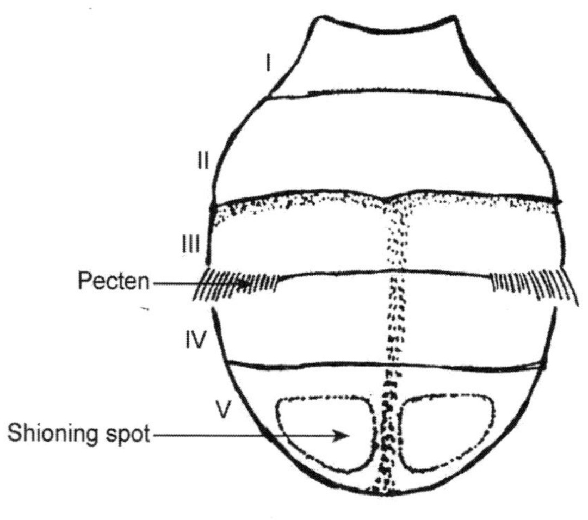

FIG. 11. Abdomen (Dorsal view)

6.3.1. Male terminalia

The terminalia are relatively small and inconspicuous and tucked away beneath the fifth tergum. In description genitalia usually refers to the visible parts of the eight and ninth segments together with the surstyli and also sometimes large proctiger.

(a) Ninth tergum : It may be oval or rounded, rather short and somewhat like a horse collar in shape having usually large membaranous proctiger posteriorly. The lower margins form long or short surstyli which offer useful generic and specific characters.

(b) Genital ring : It lies horizontally at right angle to the transverse vertical plane of the ninth tergum. It may bear a tooth like projection on each arm. It may be of various shapes and thus, sometimes, a useful generic and specific characters.

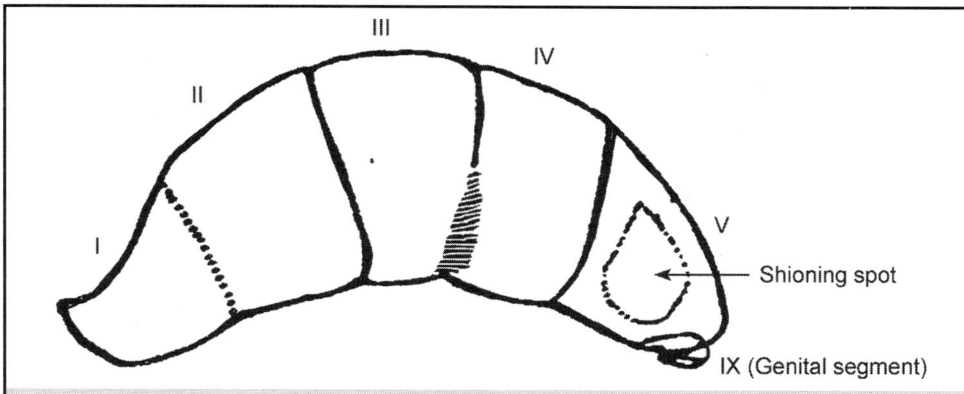

FIG. 12. Abdomen (Lateral view)

(c) Flutella : It hangs over the genital ring. It consists of a median posterior portion to which are attached a pair of lateral wings, the vanes. The vanes are straight or curved, bent or curved downwards and sometimes forwards, internal process, the phallic apodeme projects forward and upwards from the point where the median rod and vanes meet. The apodeme may be large medium sized, small or reduced to a small keel, but always flattened laterally.

(d) Phallobase : It is attached to the bifurcate posterior end of the flutella by means of a short ligament acting as a hinge.

(e) Phallotheca : It is a long coiled tube through which the seminal duct runs. It may be long or short with some part coiled. It may bear hairs or spicules, strong spines or thorns, ductules. Its length, diameter and ornamentations are important generic and specific characters.

(f) Ejaculatory apodeme : This is the pump which forces the seminal fluid through the duct to the external opening. It is flattened, the proximal part fan shaped, narrow, semi-circular or slightly to strongly asymmetrical. The stem of the apodeme is usually about half the length and narrow to rather wide. Its distal end is a membranous bulb into which leads the seminal duct. The shape and size of the apodeme are important specific characters. It shows variation according to the age of specimen.

(g) Aedeagus : It shows great diversity in form and can be very useful in segregating various genera and species. It is marked off at the end of the phallotheca by a more or less clear constriction. It may be membranous or sclerotised. In most cases, it has extra ordinary appearance : that is, with a regular arrangement of sclerotised rods or bands on the ventral aspect. A pair of curved sclerites, in various forms, is present near the base of the aedeagus.

6.3.2. Female genitalia

The post-abdomen in the female tephritid consist of the ovipositor which represent three apparent apical segments, that is the seventh, eighth and ninth. These segments are named as oviscape (basal one), inversion membrane (middle one) and aculeus (the apical one). It is normally seen that the seventh segment, that is, oviscape is the only visible part and is so referred to as ovipositor by some in the taxonomic descriptions. Some call the apical part (that is the aculeus, the ninth segment) as the ovipositor. But in actual sense the ovipositor is of three segments as mentioned above. These segments offer great taxonomic characteristics in the separation of both higher as well as lower taxa. These are as follows :

(a) Oviscape : This is a modified seventh abdominal segment bearing a pair of spiracles. It is a conical tube and acts as a protection for the apical joints which are retracted within it when not in use. In genera, it is the length of this segment which is correctly mentioned as the length of the ovipositor and is a useful taxonomic character in comparison with the length of the abdomen or the whole body. Its length may also be compared with the length of wing. It is also seen that the length of the ovipositor is approximately three times that of the oviscape. Sometimes the colour (the same or different from that of pre-abdomen) also offers a taxonomic character.

(b) Inversion membrane : It is the middle part of the ovipositor and is a membranous tube. Its surface is provided with numerous stout, regularly arranged setae which are larger on the dorsal surface. The shape and size of these, especially under high magnification offer good taxonomic character at specific levels.

(c) Aculeus : It is also called piercer or sting. It may consist of the ninth, tenth and the eleventh segments. The apex of this part especially under high magnifications offers a good taxonomic character at specific level.

6.3.3. Spermatheca

It is a small ovoid structure with a small duct opening into the genital chamber at its base. The spermatheca and its duct lie coiled up in the form of a convoluted tube at the posterior end of the eighth segment between vagina and the ventral nerve cord. It has not proved very useful in separating species as it shows variation even within a species.

6.4. COLOUR PATTERNS

The colour patterns of the wing, head, thorax and abdomen are good taxonomic characters and are used to segregate both higher and lower taxa of fruit flies.

6.4.1. Pubescens

It consists of short hair, may be sparse to dense; usually shorter on the dorsum while larger and finer on the pleura. The dorsal pubescence varies from fine and black to coarse and white or yellow; rarely very dense and reddish.

6.4.2. Wings

Wing patterns are often termed infuscations due to their fuscous colouration in Dacinie. In these fruit flies there is always a fuscous coastal band running around the coastal margin from the wing base to the apex of the wing and a fuscous streak running through the anal cell, above the anal cell extension and vein Cu1 + IA to the wing margin; the width of coastal band and other colour patterns are useful character in Dacinie species. The fruit flies of the subfamilies Trypetiane and Tephritinae are easily separated on the basis of the colour patterns of their wings. In the former the colour pattern is banded while in the latter the pattern is reticulate. Such colour patterns of wings in relation to their positions with regard to various wing cells and veins are useful generic and specific character.

6.4.3. Thorax and Abdomen

The colour of these regions vary from fuscous (mixture of black and red), fulvous (tawny or brownish-yellow), black (dull or glossy), yellow, orange brown, red-brown, brown etc. Thoracic colours are often difficult to examine under a binocular microscope due to the reflection of light off minute hairs on the mesonotum. This problem was solved by turning the specimens with the head pointing away from the light source and wetting the mesonotum with absolute alcohol using a fine hair brush.

6.5. DUST OR TOMENTUM

It is composed of microscopic short, curved, colourless, airfilled hair closely appressed to the body surface. It is quite clear when the body surface is black. The black dorsum of the thorax becomes broken and appears as being etched due to the development of even small tract of tomentum. The density of the dust varies from light and moderate to strong. Sometimes pubescene and tomentum are quite useful as generic and specific characters.

❑❑❑

7

Chapter

Distinguishing Characters of Some Important Species of Fruitfly

1. *Carpomyia vesuviana* : (Fig. 57)

(i) Chaetotaxy complete, third antennal segment typically short.

(ii) Costa without deep cleft and no lobe at the end of subcoastal cell.

(iii) Occipital setae thin, pointed, black (dark brown).

(iv) Sixth tergum of female typically shorter than fifth.

(v) Dorso-central bristles usually behind supra-alar.

(vi) Vertical surface of mesopleuran well developed.

(vii) Wings variously marked but rarely spotted.

(viii) Cubital cell drawn out into a sharp point prominent lobe at lower apex.

(ix) Section of vein Cu before joining 1A always angular.

(x) Pleuroterga bare, except for microscopic pubescence.

(xi) Four scutellar bristles.

(xii) Arista pubescent.

(xiii) Thorax and abdomen differently coloured.

(xiv) Wing variously marked.

(xv) Upper superior fronto-orbitals convergent.

(xvi) Scutellum flat.

(xvii) Wing normal in shape, variously marked.

(xviii) Wing with r-m and m cross-veins approximately parallel.

(xix) Cell first M_2 sharply pointed at apex.

2. *Bactrocera cucurbitae* : (Fig. 41)

(i) Chaetotaxy of head and thorax are reduced (ocellar, postocellar, dorso-central, presutural, sterno-pleural and usually humeral bristles absent).

(ii) Third antennal segment three times longer than wide.

(iii) Cell $1M_2$ broad, 2x wider than cell Cu and usually 2x longer than wide.

(iv) Cell Cu as long as vein Cu_1 + I^{st} A or even longer, usually with apical lobe elongated.

(v) Males with well developed stridulatory apparatus and lateral tergal glands on the fifth tergum.

(vi) Female with two spermathecae consisting of tight set coils resembling a bunch of grapes.

(vii) Wing hyaline having brown streak through the cubital area.

(viii) Abdominal terga not fused.

(ix) Posterior surstylus lobe long (at least six times as long as anterior lobe).

(x) Abdominal sternum V with slight concavity on posterior margin.

(xi) One pairs of scutellar bristles.

(xii) Pecten on abdominal tergum III of male present.

(xiii) Prescutellar bristles present.

(xiv) Male with supernumerary lobe and with a well developed stridulatory area marked with dark brown, covered with shaggy hair along vein Cu_1 + Ist A, along lower margin of cell M_4 and through cell Cu.

(xv) Three pair of inferior orbital bristles.

(xvi) Coastal band broad, apical cell extending in cell $R_{3.}$

(xvii) A broad brown mark over r-m.

(xviii) Scutellum yellow.

3. *Bactrocera dorsalis* : (Fig. 28)

(i) Chaetotaxy of head and thorax are reduced (ocellar, postocellar, dorso-central, presutural, sterno-pleural and usually humeral bristles absent).

(ii) Third antennal segment three times longer than wide.

(iii) Cell $1M_2$ broad, 2x wider than cell Cu and usually 2x longer than wide.

(iv) Cell Cu as long as vein Cu_1 + I^{st} A or even longer, usually with apical lobe elongated.

(v) Males with well developed stridulatory apparatus and lateral tergal glands on the fifth tergum.

(vi) Female with two spermathecae consisting of tight set coils resembling a bunch of grapes.

(vii) Wing hyaline having brown streak through the cubital area.

(viii) Abdominal terga not fused.

(ix) Posterior surstylus lobe long (twice as long as anterior lobe).

(x) Abdominal sternum V of male with a deep concavity on posterior margin.

(xi) Supra-alar bristles present.

(xii) Mesonotum with post sutural yellow vittae.

(xiii) Head usually yellow with black markings.

(xiv) Coastal band continuous at least as narrow band to wing apex.

(xv) Coastal band slightly expanded.

(xvi) Femora largely or entirely yellow, not more than tinges of brown at apices.

(xvii) Terga three to five fulvous, with a transverse fuscous to black band along anterior margin of tergite III and a medial longitudinal fuscous to black band on tergite three to five.

(xviii) Coastal band narrow, complete, no brown spot at apex of Cu1+ 1A.

(xix) Median longitudinal vita from tergite three to end of fifth.

(xx) Coastal band extending along below R_{2+3} except at apex of wing.

4. *Bactrocera correcta* : (Fig. 30)

(i) Chaetotaxy of head and thorax are reduced (ocellar, postocellar, dorso-central, presutural, sterno-pleural and usually humeral bristles absent).

(ii) Third antennal segment three times longer than wide.

(iii) Cell $1M_2$ broad, 2x wider than cell Cu and usually 2x longer than wide.

(iv) Cell Cu as long as vein Cu_1 + I^{st} A or even longer, usually with apical lobe elongated.

(v) Males with well developed stridulatory apparatus and lateral tergal glands on the fifth tergum.

(vi) Female with two spermathecae consisting of tight set coils resembling a bunch of grapes.

(vii) Wing hyaline having brown streak through the cubital area.

(viii) Abdominal terga not fused.

(ix) Posterior surstylus lobe long (twice as long as anterior lobe).

(x) Abdominal sternum V of male with a deep concavity on posterior margin.

(xi) Supra-alar bristles present.

(xii) Mesonotum with post sutural yellow vittae.

(xiii) Head usually yellow with black markings.

(xiv) Coastal band interrupted in cell R_3 (beyond tip of vein R_{2+3} or lacking beyond cell Sc).

(xv) Face with black spot, one in each antennal furrow.

(xvi) Mesonotum with rather indistinct submedian narrow black vittae.

(xvii) Scutellum yellow.

(xviii) Hind tibia with a prominent keel-like process on posterior dorsal surface before apex.

(xix) Abdomen rufous.

(xx) No cubital streak.

5. *Bactrocera zonata* : (Fig. 29)

(i) Chaetotaxy of head and thorax are reduced (ocellar, postocellar, dorso-central, presutural, sterno-pleural and usually humeral bristles absent).

(ii) Third antennal segment three times longer than wide.

(iii) Cell $1M_2$ broad, 2x wider than cell Cu and usually 2x longer than wide.

(iv) Cell Cu as long as vein Cu_1 + I^{st} A or even longer, usually with apical lobe elongated.

(v) Males with well developed stridulatory apparatus and lateral tergal glands on the fifth tergum.

(vi) Female with two spermathecae consisting of tight set coils resembling a bunch of grapes.

(vii) Wing hyaline having brown streak through the cubital area.

(viii) Abdominal terga not fused.

(ix) Posterior surstylus lobe long (twice as long as anterior lobe).

(x) Abdominal sternum V of male with a deep concavity on posterior margin.

(xi) Supra-alar bristles present.

(xii) Mesonotum with post sutural yellow vittae.

(xiii) Head usually yellow with black markings.

(xiv) Coastal band interrupted in cell R_3 (beyond tip of vein R_{2+3} or lacking beyond cell Sc).

(xv) Face with one cross band furrow.

(xvi) Prescutellars yellow.

TABLE 5 : Fruit flies and their hosts

	Fruit fly species	*Major Hosts*
1.	*Carpomyia vesuviana*	Ber (*Ziziphus mauritiana*)
2.	*Bactrocera cucurbitae***	Bitter gourd (*Momordica charantia*), sponge gourd (*Luffa acutangula*) and all cucurbits
3.	*Bactrocera dorsalis**	Mango (*Mangifera indica*),Guava (*Psidium guajava*)
4.	*Bactrocera correcta**	Guava (*Psidium guajava*)
5.	*Bactrocera zonata**	Cucumber (Cucumis sativus), Guava (*Psidium guajava*)

** Attracted to Cue-lure

* Attracted to methyl eugenol

Key to taxa

Efforts have been made to prepare a key for the Indian species especially predominant in northern India. The characters have been compiled in a best possible way to identify the five fruit fly species which have been identified on the basis of male fruit fly species obtained from rearing, damaged fruits and vegetables or attracted to attractants.

It is worth mentioning here that the five species which are identified have been obtained from the host. A guide for identification of these fruit flies is prepared which can be a ready, rackoner for concerned workers and farmers. The key on the sub-family Dacinae is based on the description made by Drew (1989).

1. Chaetotaxy of head and thorax are reduced (ocellar, postocellar, dorso-central, presutural, sterno-pleural and usually humeral bristles absent); third antennal segment three times longer than wide; abdominal terga not fused; wing hyaline having brown streak through cubital area. (Fig. 13) .... genus *Bactrocera* (2).

   Chaetotaxy complete; third antennal segment typically short, arista pubescent; occipital setae thin pointed; dorsocentral bristles behind the supra-alars; wings variously marked but rarely spotted, cubital cell drawn out into a prominent point at lower apex, four scutellar bristles, scutellum flat with five black marks (Fig. 21, 22 and 23) .... *Carpomyia vesuviana.*

2. Posterior surstylus lobe long (at most twice as long as anterior lobe); abdominal sternum V of male with a deep concavity on posterior margin; Median vita absent; Supra-alar bristles present; Mesonotum with two post sutural yellow vittae; Head yellow with black marking (Fig.15)........................(3).

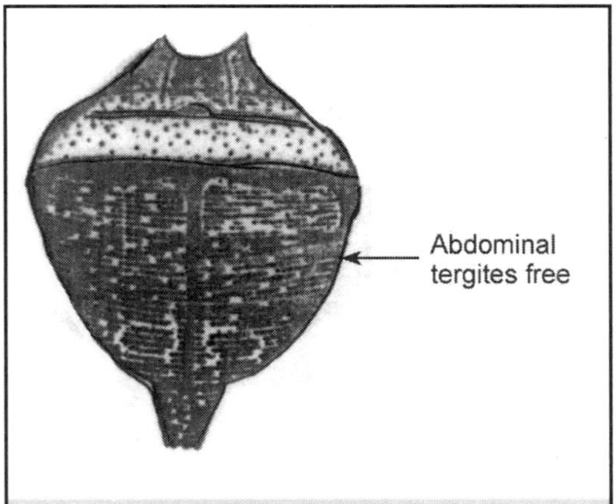

FIG. 13. Abdomen (*Bactrocera*)

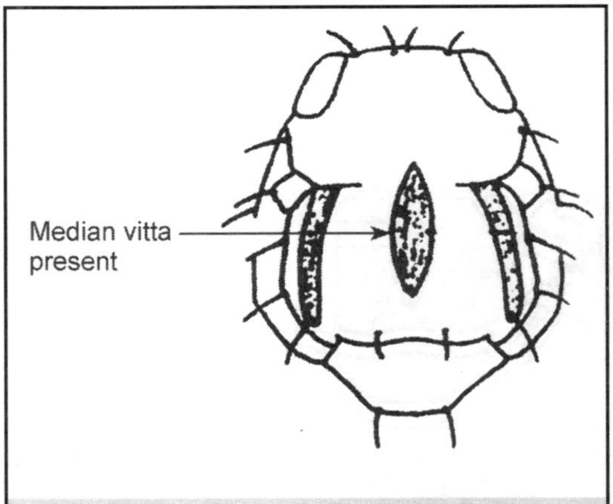

FIG. 14. Thorax (*B. cucurbitac*)

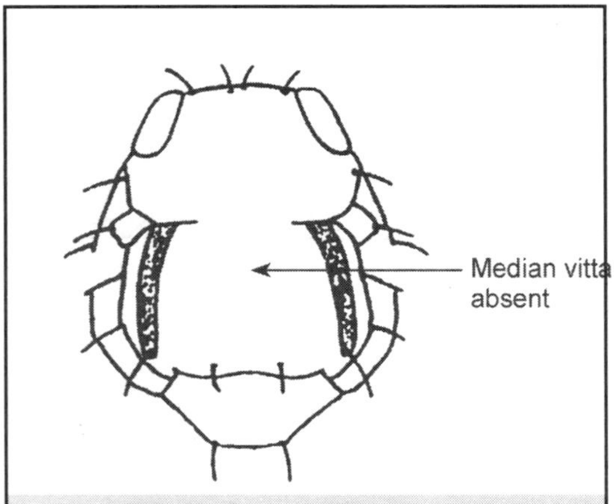

FIG. 15. Thorax (*B. correcta*)

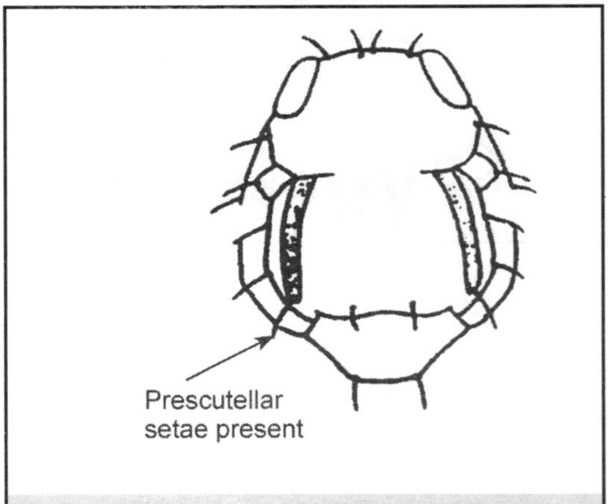

FIG. 16. Thorax (*B. zonata*)

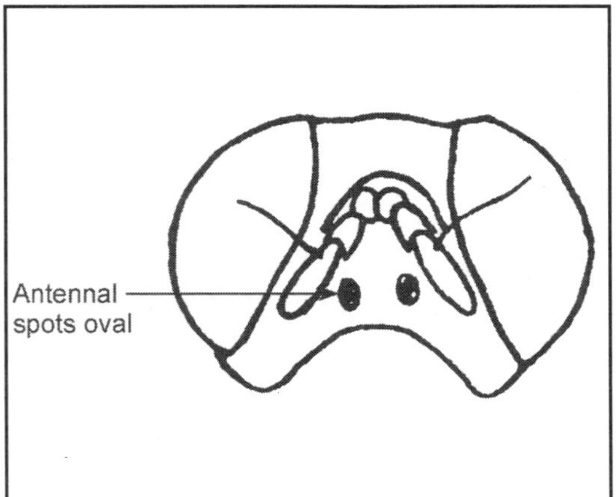

FIG. 17. Head (*B. zonata*)

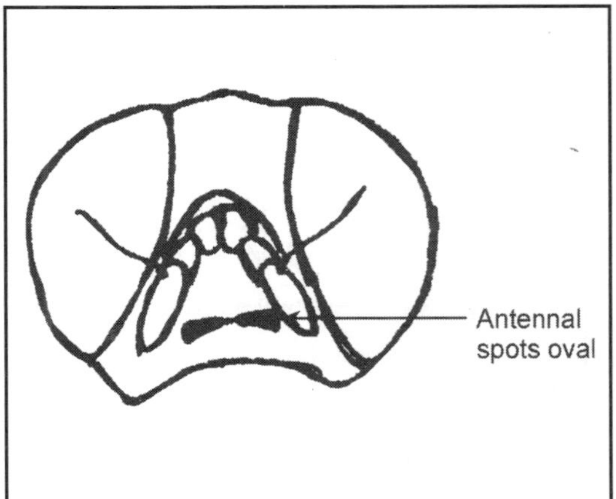

FIG. 18. Head (*B. currecta*)

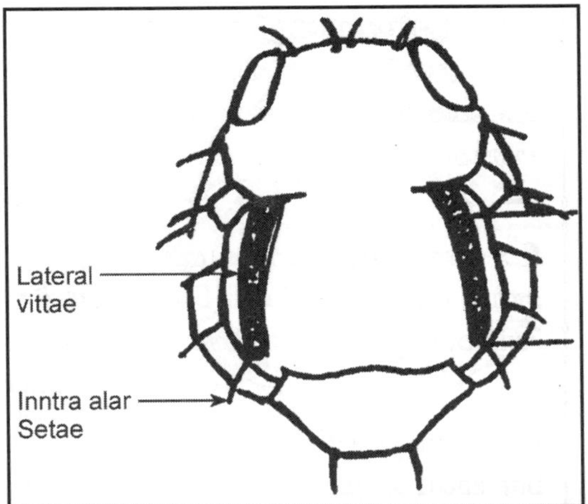

FIG. 19. Thorax (*B. dorsalis*)

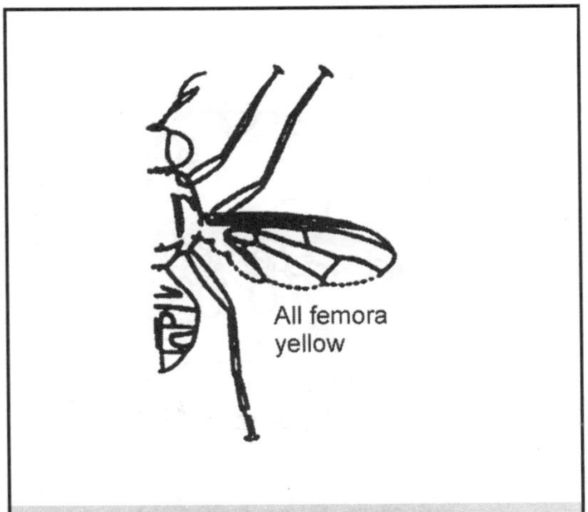

FIG. 20. Leg (*B. dorsalis*)

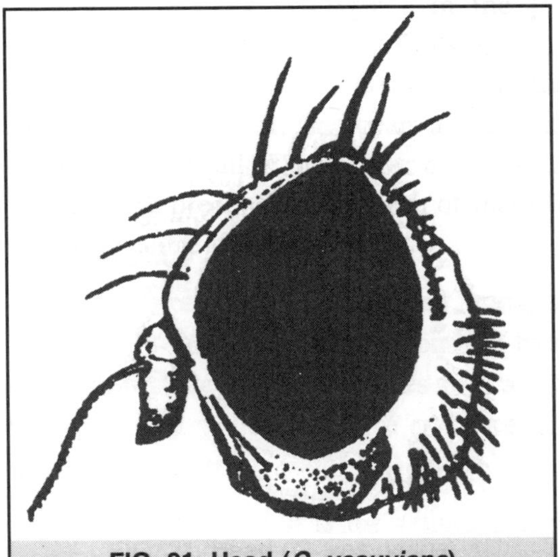

FIG. 21. Head (*C. vesuviana*)

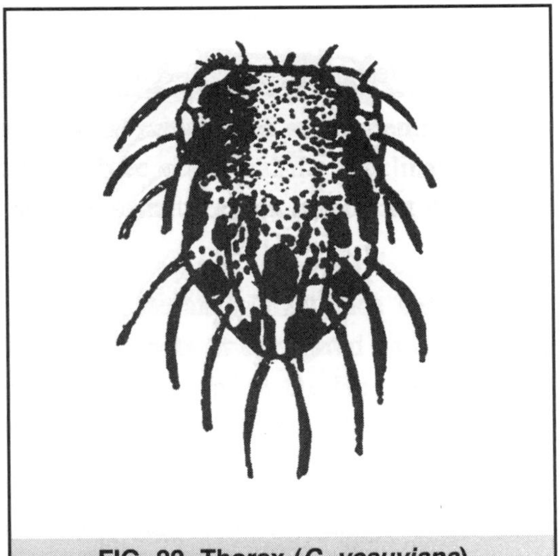

FIG. 22. Thorax (*C. vesuviana*)

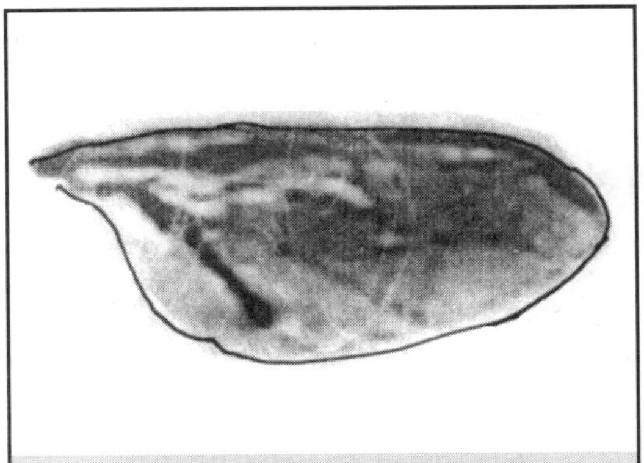

FIG. 23. Wing (*C. vesuviana*)

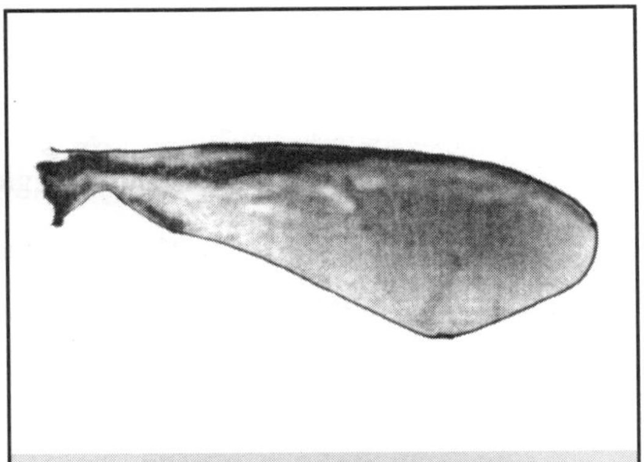

FIG. 24. Wing (*B. dorsalis*)

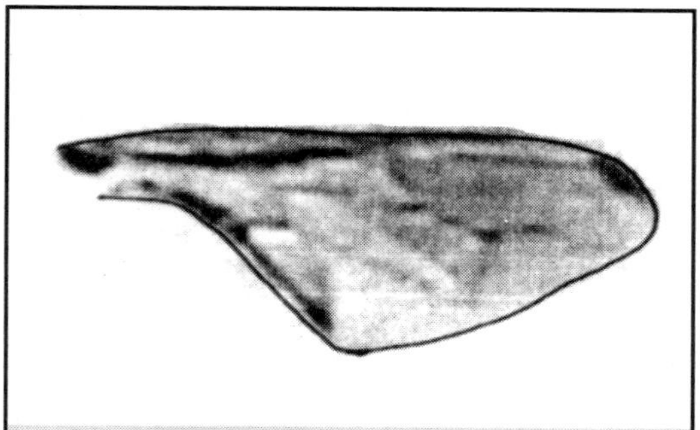

FIG. 25. Wing (*B. correcta*)

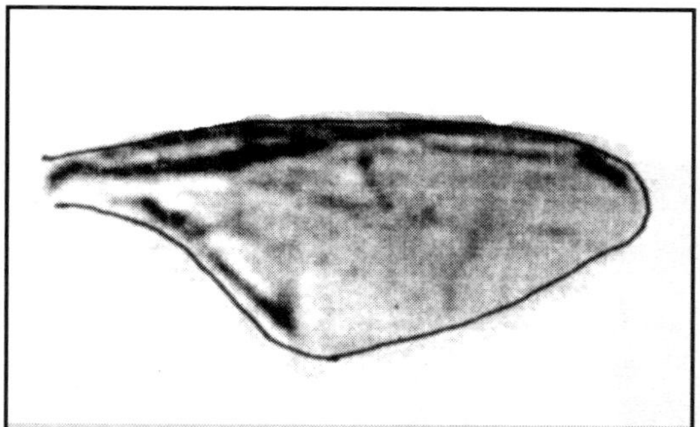

FIG. 26. Wing (*B. zonata*)

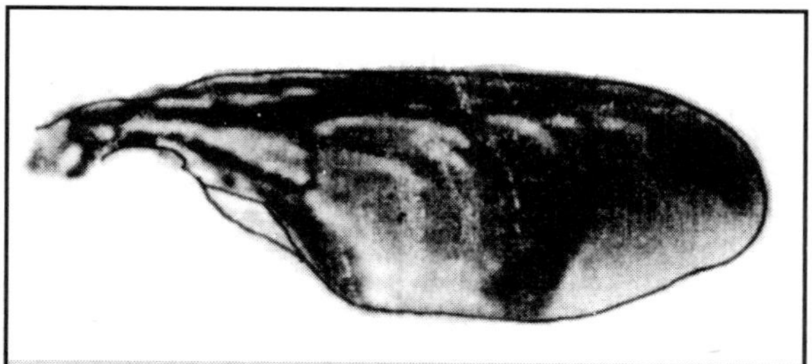

FIG. 27. Wing (*B. cucurbitae*)

— Posterior surstylus lobe long (at least six times as long as anterior lobe); abdominal sternum V with slight concavity on posterior margin; one pairs of scutellar bristles, antennal spots oval vein dm-cu infuscated antennal spots present, median vittae present; male with a well develop stridulatory area marked with dark brown; coastal band extending along below R_{2+3} except at apex of wing (Fig. 14, 17 and 27) .... *Bactrocera cucurbitae.*

3. Coastal band complete as narrow band to wing apex; femora largely, entirely yellow, not more than tinges of brown at apices; antennal spots oval; lateral vittae ending after intra –alar setae. (Fig. 19, 20, 24) .... *Bactrocera dorsalis.*

— Coastal band interrupted in cell R_2 (beyond tip of vein R_{2+3} or lacking cell Sc. (Fig. 25 and 26) .... (4).

4. Face with black spot one in each antennal furrow; antennal spots fuse; thorax black, scutellum yellow; cell Cu as long as vein Cu + I^{st} A with apical lobe elongated, abdomen rufous, no cubital streak. (Fig. 18 and 25) .... *Bactrocera correcta.*

— Face with one cross band in furrow; Prescutellars present; antennal spots oval; thorax reddish brown. (Fig. 16 and 17) .... *Bactrocera zonata.*

❑❑❑

8

Chapter

Species of Economic Importance

Of the 4500 known species of fruit fly worldwide, nearly 200 are considered as pests but about 70 species are regarded as agriculturally important throughout the world (Clark *et al.,* 2005). It is reported that only five percent of all Tephritid species are economically important. The tribe dacini with genus *Bactrocera* is of importance in India and from the economic point of view, oriental fruit fly or mango fruit fly *Bactrocera dorsalis* (Hendel), guava fruit fly *Bactrocera correcta* (Bazzi) and peach fruit fly *Bactrocera zonata* (Saunders) are very important pest of fruit crops and are recognized worldwide as the most important threat to horticulture (Verghese *et al.,* 2004, Kapoor 2005, Sharma *et al.,* 2011). David and Ramani (2011) reported that 325 species of fruit flies are known to occur in Indian sub-continent of which 243 in 79 genera are from India alone.

According to (Kapoor, 2005), there are as many as 205 fruit fly species occurring in India but major pest species which have been recorded from horticultural crops belong to five genera. The important tephritid species of India with their hosts, distribution, taxonomy and biology are summerised under two Sub families :

Sub Family : Dacinae

(a) Genus : *Bactrocera*

(b) Genus : *Dacus*

Sub Family : Tephritinae

(c) Genus : Cyrtostola

(d) Genus : *Pliomelaena*

(e) Genus : *Dioxyna*

Major Pests

1. *Bactrocera (B.) dorsalis* (Hendel)
2. *Bactrocera (B.) zonata* (Saunders)
3. *Bactrocera (B.) correcta* (Bezzi)
4. *Bactrocera (B.) carambole* Drew and Hancock
5. *Bactrocera (B.) caryeae* (Kapoor)
6. *Bactrocera (B.) vishnu* Drew and Hancock
7. *Bactrocera (B.) latifrons* (Hendel)
8. Bactrocera (B.) *affinis* (Hardy)
9. *Bactrocera (B.) nigrotibialis*
10. *Bactrocera (B.) nigrofemoralis* (White and Tsuruta)
11. *Bactrocera (B.) verbascifoliae*
12. *Bactrocera (B.) versicolor* (Bezzi)
13. *Bactrocera (B) paraverbascifoliae* (Drew and Raghu)
14. *Bactrocera (Zeugodacus) cucurbitae* (Coqillett)
15. *Bactrocera (Z.) caudata* (Fabricious)
16. *Bactrocera (Z.) tau* (Walker)
17. *Bactrocera (Z.) scutellaris* (Bezzi)
18. *Bactrocera (Z.) oleae* Gmelin
19. *Bactrocera (Z.) gavisa*
20. *Bactrocera (Hemigymnodacus) diversa (Coquillet)*
21. *Bactrocera (Javadacus) trilineata* Hardey
22. *Dacus (Didacus) ciliatus* Loew
23. *Dacus (Callantra) ramanii*
24. *Dacus (Callantra) longicornis* Weidemann
25. *Dacus (Callantra) sphaeroidalis* (Bezzi)
26. *Cyrtostola limbata* (Hendel)
27. *Pliomelaena udhanpurensis* Agarwal and Kapoor
28. *Dioxyna sororcula* (Weidemann)

Minor Pests

29. *Carpomyia vesuviana* Costa
30. *Carpomyia pardalina* (Bigot)
31. *Acanthiophilus helanthi* (Rossi)

(A) GENUS : *BACTROCERA*

Subgenus : *Bactrocera* Macquart

8.1. *BACTROCERA* (*BACTROCERA*) *DORSALIS* HENDEL (ORIENTAL FRUIT FLY)

The most notorious pest of mango and other fruits, *Bactrocera dorsalis* has been known in old literature as *Dacus ferrugineus* (Fabricius). This name preoccupied was later described by Dr. D.E. Hardey of University of Hawaii as *Dacus dorsalis*. He regarded *Dacus* as *Bactrocera* for most of the pest species. This is popularly known as mango fruit fly in India and oriental fruit fly the world over. It attacks 89 species of fruits including mango in Asia pacific region.

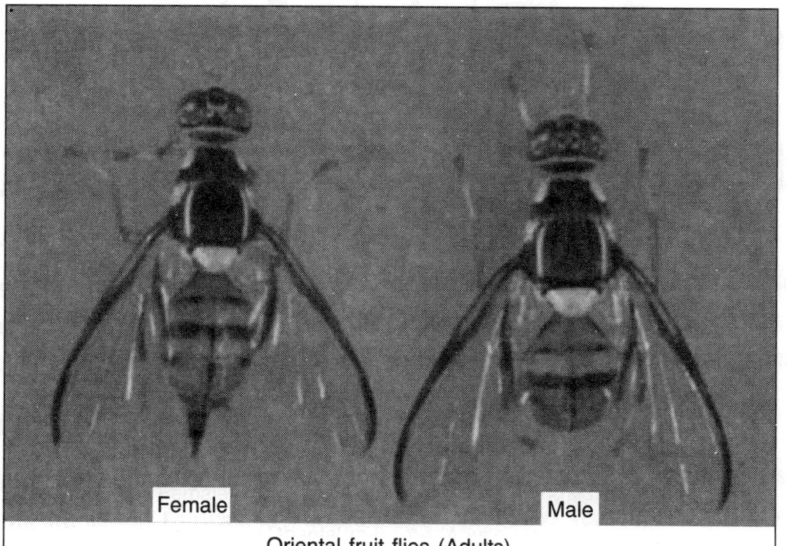

Oriental fruit flies (Adults)

FIG. 28. *Bactrocera* (*B.*) *dorsalis* (Hendel)

8.1.1. Geographical Distribution

B. dorsalis occurs in the northern part of the range of the *dorsalis* complex in Asia, exclusively in the Northern Hemisphere (with the marginal exception of Nauru). The other species occur in the south of the Indian subcontinent and in South East Asia.

Apart from India where it has assumed the status of a major pest, are Pakistan,

Nepal, Bhutan, Taiwan, Thailand, Laos, Vietnam, Sri Lanka, Myanmar, Malaysia, Indonesia, Mauritius, Phillipines, and Cambodia in Asia. Also it is found in East Indies, China, Formosa, Bonin Islands, Hawaii and Australia.

It is widely distributed in the Oriental region hence it is also called Oriental fruit fly. The pest is active throughout the year in south India, whereas, in northern parts, the pest hibernates during winters (November to March) in pupal stage.

8.1.2. Host range

The host plant of *B. dorsalis* include mango, guava, loquat, apricot, plum,fig, apple, peach, pear, quince, persimmon, banana, alubukhara, pomegranate, chiku, jujube, sweet lime, malta orange, santra, pomelo, sour orange, orange, jack fruits, bael and brinjal. The indices of infestation vary in different fruits. Some of the hosts are used only as ovipositional sites (Narayanan and Batra 1960). In Hawaii it has been recorded from over 250 host fruits. *B. dorsalis* was also collected first time from the islands of North, Middle and South Andaman on guava, mango, *Terminalia procera, T. manii, Artocarpus* sp. and *Ayzygium* sp. in South America.

8.1.3. Nature and symptoms of damage

The adult female lays eggs by making a deep puncture by its ovipositor into the skin of the fruit. This puncture is inconspicuous as its colour blends with the dark green colour of the fruit. In some yellow and pale brown varieties it is very clearly visible. The maggots on hatching feed on the pulp of the fruit for few days and a brown rotten patch appears on the fruit surface. The mesocarp becomes dirty brown. Infested fruits finally fall on the ground and the maggots come out from these fallen fruits to pupate in soil. Semi ripe fruits are attacked usually by April-May.

The seasonal activity was recorded highest during July in Punjab (Mann 1996) while in northern Karnataka the peak catches were recorded during 21st, 23rd and 46th standard weeks (May-June and October) (Babu and Virakthamath 2003). The *B. dorsalis* population did not show any significant correlation with weather parameters. In Andhra Pradesh *Bactrocera* spp. was recorded maximum on mango during May-July coinciding with fruit maturity period (Sarada *et al.,* 2001). In Himanchal Pradesh maximum mixed population of *B. dorsalis* and *B. zonata* was recorded during July in mango orchards while in guava the highest population was recorded in September (37th and 39th standard week) (Gupta and Bhatia, 2000). In later studies (Ranjitha and Viraktamath 2005), in Karnataka, *B. dorsalis* peaked on 30th standard week. *B. dorsalis* was recorded as the dominant fruit fly among the 4 species recorded (Babu and Virakthamat, 2003). A significant correlation with minimum tepmperrature and wind speed was observed in Karnataka (Verghese and Sudhadevi, 1998) against *B. dorsalis*.

In China the peak period of adult activity was during June-July, with the exception of citrus and sweet orange orchard where it was September-October and October-December, respectively (Zang *et al.,* 1995)

8.1.4. Biology

B. dorsalis will not develop at temperatures below 13°C. Pupariation is in the soil under the host plant and adults emerge after 1-2 weeks (longer in cool conditions) and adults occur throughout the year (Christenson and Foote, 1960). *B. dorsalis* is a tropical species which would be unable to survive the winter in the EPPO region, except possibly in the south. The adults are best able to survive low temperatures, with a normal torpor threshold of 7°C, dropping as low as 2°C in winter. Though little is specifically known about the biology of the other pest species of the *B. dorsalis* complex, there is no reason to suppose that it differs greatly from that of *B. dorsalis*; if anything, they could be supposed from their distribution to be even more specifically adapted to tropical conditions.

Female flies insert eggs under the skin of fruit in clusters of 2 to 15 below the fruit surface. The eggs measure 1/25 by 1/250 inch and are white, elongate and elliptical. The larvae are white, legless like elongated cone. The mouth is at the pointed end of the body. There are three larval stages or instars. The third larval instar is about 2/5″ inch long. The entire larval stage lasts for 11-15 days. The larvae when mature drop to the ground and pupate in the soil. Puparium is yellowish brown and seed like. Adults emerge in about 10 days.

8.1.5. Oviposition

The adult males and females usually mate at dusk and mating lasts for an hour or more. The pre oviposition period may be two to fifteen days depending upon climatic conditions and diet. The females lay eggs singly or in clusters, mostly into the pulp of the developing fruit, just underneath the skin of the fruit. Tactile bristles on the ovipositor assist the fly to discriminate between hard and soft surface for oviposition. The female, after finding a suitable host bends its abdomen at right angle to the axis of the body and the distal, needle like ovipositor is inserted deep into the fruit till it disappears. The eggs are then deposited and the ovipositor is withdrawn, leaving a puncture. The cluster may contain 2 to 15 eggs depending upon the softness of fruit pulp. In one month the female may lay 150-200 eggs. The incubation period of eggs may be two to three days during March-April, but during summers it may last 24-36 hours and may be 10 days in winters.

Oviposition behaviour of *B. dorsalis* on mango variety ‘Malika’ showed that infestation began as early as 5 weeks before fruit maturity while in ‘Amrapali’ infestation began just 2 weeks prior to fruit maturity. (Kalia and Srivastava, 1992).

Egg : Eggs of *B. dorsalis* are laid below the skin of the host fruit. These hatch within 1-3 days and the larvae feed for another 9-35 days.

Larva : The young larva is white, translucent and measures 1.5 mm × 0.3 mm. the full grown larva is 8-9 mm long and 1.5 mm broad. The colour of larva is yellowish and opaque. The head cephalic segment is pointed anteriorly and trapezoidal in outline. It bears on its dorsal surface two antennal protuberances and maxillary palps. Posterior spiracles are very conspicuous present on last abdominal segment. The oral ridges consist of 11-15 rows of short teeth which are less than 0.25 mm deep. A stigmata or stomal sensory organ are found. The pre oral ridges are without serrated edges. Posterior spiracular hairs with dorsal and ventral bundles of 17-20 hairs and lobes are surrounded by small fine spinules. A pair of anterior spiracles is found on the first thoracic segment. The larval period lasts for 6-9 days. There are three instars. The last instar larva has the habit of jumping a short distance to find a suitable place for pupation.

Among the mango varieties, the survival percentage was highest on 'Bangalora' and lowest in 'Amrapali'. However, among guava, *B. dorsalis* preferred guava variety 'Allahabad safeda' (Kalia, 1992).

Pupa : Pre pupal period ranges 18 hours to 2 days. Pupa is cylindrical in shape. The size is 4-5 mm × 2 mm. The colour varies from reddish yellow to ochraceous. The Pupa is divided into 11 segments, the last being more prominent. The anterior spiracles are similar to larva but are much darker. Posterior spiracles are situated below the mid-horizontal line which are yellowish. Each spiracle has three small but relatively broad openings. Pupal period lasts for 6-44 days. It normally takes place three to seven inches below soil surface. High humidity is required for the development of the adult.

Adult : The adult fly is highly variable but mostly yellow with dark markings on the thorax and abdomen. Generally, the abdomen has two horizontal black stripes and a longitudinal median stripe extending from the base of the third segment to the apex of the abdomen. These markings may form a "T" shaped pattern but the pattern varies considerably.

The adult measures 14 mm across the wings and 7 mm in length. Male is slightly smaller than the female. Female possesses a tapering abdomen which ends in pointed ovipositor. The fly is brown or dark brown in colour with hyaline wings and yellow legs. Thorax is ferruginous.

The face of the adult is yellow with circular black spots. Scutum is black, lateral post sutural vittae is broad and end beyond intra-alars. Costal band of the wing is confluent with R_{2+3}. The legs are fulvous with rows of long setae on fore femora. The abdomen is orange to dark brown with prominent T-shaped mark.

In male flies the third abdominal tergum possess pecten. These are attracted to methyl eugenol.

8.1.6. Morphology

Drew and Hancock (1994) described *B. dorsalis* in detail. Recently, DNA probes have been proposed as a practical means of discriminating between all life stages of the three main tephritids present in Hawaii (*C. capitata*, *Bactrocera cucurbitae* and *B. dorsalis*) (Haymer *et al.*, 1994).

External Morphology : Face marked with a dark spot in each antennal furrow; scutum predominantly black, except for lateral yellow vittae, and yellow postpronotal lobe and notopleurae; scutellum entirely pale-coloured, except sometimes for a narrow black line across the base; costal margin of wing with a distinct coloured band from the end of vein Sc to just beyond the end of vein R_{4+5}; crossveins r-m and dm-cu not covered by any markings; abdominal tergites three to five with a distinct black 'T'-shaped mark; postpronotal lobes without any setae (sometimes with some small setulae or hairs).

Head : With reduced chaetotaxy, lacking ocellar and postocellar setae; first flagellomere at least three times as long as broad.

Thorax : With reduced chaetotaxy, lacking dorsocentral and katepisternal setae. Scutum with anterior supra-alar setae and prescutellar acrostichal setae; scutellum not bilobed, with only two marginal setae (the apical pair).

Wing : Vein Sc abruptly bent forward at nearly 90°, weakened beyond this bend and ending at subcostal break; vein R1 with dorsal setulae; cell cup very narrow, about half depth of cell bm; cup extension very long, equal or longer than length of vein A1 + CuA2. 5-7 mm long. Cell bc and basal half of cell c without a complete covering of microtrichia.

Abdomen : All tergites separate (view from side to see overlapping sclerites); tergite five with a pair of slightly depressed areas (ceromata). Male with a row of setae (the pecten) on each side of tergite three.

8.1.7. Means of Movement and Dispersal

Adult flight and the transport of infested fruits are the main means of movement and dispersal to previously uninfested areas. Numerous fruit species on which *B. dorsalis sensu lato* was intercepted in Japan in the 1980s, *B. dorsalis sensu lato* was by far the most frequently intercepted tephritid (80%) on prohibited fruit in a study at Osaka Airport, Japan. Many *Bactrocera* spp. can fly 50-100 km (Fletcher, 1989). Some host fruits are only infested when ripe, and this has been the basis for an "infestation-free quarantine procedure" for avocados exported from Hawaii to mainland USA, which was recently called into question when fruits still on the tree were found to be infested.

8.1.8. Pest Significance

Economic impact : *B. dorsalis* is a serious pest of a wide variety of unrelated fruit crops, but seldom cucurbits. The most notorious pest of mango and other fruits, *Bactrocera dorsalis* has been known in old literature as *Dacus ferrugineus* (Fabricius). This name preoccupied was later described by Dr. D.E. Hardey of University of Hawaii as *Dacus dorsalis*. Wood (1989) regarded *Dacus* as *Bactrocera* for most of the pest species. This is popularly known as mango fruit fly in India and oriental fruit fly the world over. It attacks 89 species of fruits including mango in Asia pacific region.

It is one of the five most important pests of agriculture in South East Asia. On this basis, at least the more widely distributed of the other pest species of the complex (*B. carambolae*, *B. papayae*) are probably as damaging. It should be noted, however, that Drew and Hancock (1994) attribute pest status to the group of species mentioned in his data sheet without specific supporting information. I.M. White (personal comm.) reports that these two species and *B. occipitalis*, *B. philippinensis* and *B. kandiensis* are as damaging as *B. dorsalis*, while *B. caryeae* is probably as damaging and *B. pyrifoliae* is only a pest in a very limited area.

8.1.9. Phytosanitary Risk

EPPO lists *B. dorsalis* as an A1 quarantine pest within the broad category "non-European Trypetidae"; it is also of quarantine significance to APPPC, COSAVE, CPPC, IAPSC, JUNAC and OIRSA. *B. dorsalis* is indigenous to Asia, but like other *Bactrocera* spp. is known by experience to have the potential to establish adventive populations in various other tropical areas. Its presence in Hawaii, but not in mainland USA, has contributed to its high international profile as a quarantine pest. Occurring in northern India and in parts of central China, *B. dorsalis* has a geographical range which is less tropical than the other members of the complex. The direct risk of establishment of *B. dorsalis* in most of the EPPO region is minimal, though populations might enter and multiply during the summer months. In southern areas, such populations could possibly survive one or several winters; it is difficult to judge how directly damaging they might be. The major risk for EPPO countries arises from the probable imposition of much stricter phytosanitary restrictions on exported fruits (particularly to America) if *B. dorsalis* enters and multiplies, even temporarily.

8.1.10. Phytosanitary Measures

Consignments of fruits of especially *Annona*, *Averrhoa carambola*, *Citrus*, *Fortunella*, *Malus*, *Mangifera indica*, *Prunus domestica*, *Prunus persica*, *Psidium guajava* and *Pyrus* from countries where *B. dorsalis* occurs should be inspected for symptoms of infestation and those suspected should be cut open in order to look for larvae. EPPO recommends that such fruits should come from an area where *B. dorsalis* does not occur, or from a place of production found free from the pest by regular inspection for 3 months before harvest. Fruits may also be treated in transit by cold treatment (*e.g.* 11, 12 or 14 days at 0.5, 1.0 or 1.5°C, respectively or 19, 25 or 25 days at 5, 6 or 7°C, respectively, for temperature-sensitive fruits like mangosteens; (Burikam *et al.*, 1992) or, for certain types of fruits, by vapour heat (*e.g.* keeping at 43°C for 4-6 h) (USDA, 1994), hot-water treatment (46°C for 65-90 min, according to size and shape of fruits), or forced hot-air treatment (Armstrong *et al.*, 1995). Ethylene dibromide was previously widely used as a fumigant but is now generally withdrawn because of its carcinogenicity; methyl bromide is less satisfactory, damaging many fruits and reducing their shelf life, although treatment schedules are available for specific cases (*e.g.* for cucumbers at 21-26°C, 32 g/m^3 for 2 h). Until the true pest status of the potential pest species of the *dorsalis* complex has been established, it would be reasonable to continue making these requirements for consignments from areas where any of them occur.

Plants of host species transported with roots from countries where *B. dorsalis* occurs should be free from soil, or the soil should be treated against puparia. The plants should not carry fruits. Such plants may indeed be prohibited importation.

8.1.11. Detection and Identification

Attacked fruit will usually show signs of oviposition punctures. Fruit with high sugar content, such as peaches, will exude a sugary liquid, which usually solidifies adjacent to the oviposition site.

8.1.12. Control

When detected, it is important to gather all fallen and infested host fruits, and destroy them. Those species whose males are attracted to lures should be continually monitored using bait traps (Bateman, 1982). Insecticidal protection is possible by using a cover spray or a bait spray. Malathion is the usual choice of insecticide for fruit fly control and this is usually combined with protein hydrolysate to form a bait spray; practical details are given by Bateman (1982). Bait sprays work on the principle that both male and female tephritids are strongly attracted to a protein source from which ammonia emanates. Bait sprays have the advantage over cover sprays in that they can be applied as a spot treatment so that the flies are attracted to the insecticide and there is minimal impact on natural enemies. Biological control has been tried against *B. dorsalis sensu lato*, but introduced parasitoids have had little impact. Male annihilation, utilizing the attraction of males to methyl eugenol was used to eradicate *B. dorsalis* from the northern Ryukyu Islands, Japan (Cunningham, 1989b). The sterile insect technique (SIT), requiring the release of millions of sterile flies into the wild population so that there is a strong likelihood of wild females mating with sterile males (Gilmore, 1989), was used to eradicate *B. dorsalis* from the Ogasawara Islands, Japan.

8.2. *BACTROCERA ZONATA* (SAUNDERS)

Bactocera zonata is commonly known as peach fruit fly. Since it was more often recovered from peach fruit therefore the name peach fruit fly. This fly is in abundance in fruit orchards and is next to *B. dorsalis* in respect of severity of damage in mango fruits. Although it is the major pest of peach but it infests other fruits in serious form.

8.2.1. Distribution

This is originally an Indian species, first reported from Bengal. Presently it is widely distributed in Pakistan, Sri Lanka, Nepal Bangladesh, Egypt and southeast Asia.

8.2.2. Host range

A variety of fruits recorded are peach (*Prunus persicae*), mango (*Mangifera indica*), fig (*Ficus carica*), guava (*Psidium guajava*), sapota (*Achros sapota*), ber (*Zizyphus jujube*), citrus (*Citrus* sp.), bael (*Aegle marmelose*), tomato (*Lycopersicon escutenlum*), custard apple (*Anona squamosa*), pomegranate (*Punica granatum*), apple (*Pyrus malies*) and pear (*Pyrus communis*).

The species attacks a number of fruits and vegetables such as fig (*Ficus careica*), mango (*Mangifera indica*), chiku (*Manilkara achras*), Ber (*Zizyphus mauritiana* Lam), guava, sapota, citrus, bael, bottle gourd, tomato, long melon, *Luffa*, *Careya arborea*, brinjal, Mahua (*Madhuca indica*), pomegranate (*Punica granatum*), pear (*Pyrus communis*), *Basolina latifolia*, custard apple and apple.

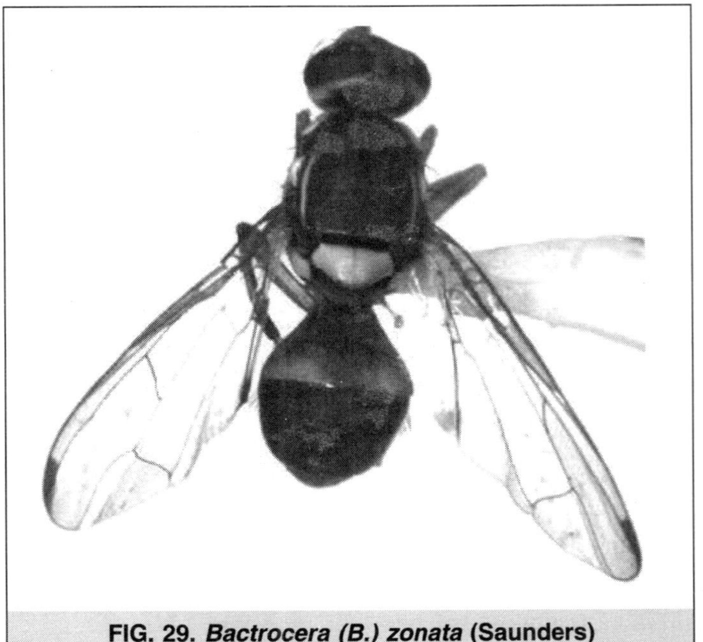

FIG. 29. ***Bactrocera (B.) zonata*** **(Saunders)**

8.2.3. Nature of damage

The fruits show sign of light coloured puncture due to ovipostion by the female. Sometimes oozing fluid and then rotting fruit ultimately falls down. Fruits with high sugar content such as peaches exude a sugary liquid which solidifies adjacent to the oviposition site.

8.2.4. Biology

The flies remain active throughout the year except in severe winter months *i.e.* January to February. The life history of *B. zonata* resembles that of *B. dorsalis*. The pre-ovipostion period is about 20 days. The female lays about 137 eggs in its life span. The eggs are laid under the rind of fruit in clusters of two to nine. The fully grown maggot falls on the ground and pupates in soil.

The pest over winters in the pupal stage. The adult flies also congregate under the leaves and bushes during winter months like *B. dorsalis*. In spring whatever hosts are available, it lays eggs on it. It attacks loquat and peach in May and June. In the later moths it transfers to cucurbits which are abundant during summers. During rainy season it passés its life cycle on mangoes, guava, chiku, pomegranate, citrus bael etc. At this time the population is at its peak. The fly population was positively correlated with minimum temperature and rainfall and negatively correlated with relative humidity (Sarada *et al.,* 2001).

In Pakistan average season of fruit infestation by *B. zonata*, and *B. cucurbitae* in melon was 29% and in guava 44% and 9% in mango in unprotected fields. (Stonehouse *et al.,* 2001). The seasonal activity of fruit flies showed that the population of *Bactocera zonata* was higher than that of *B. dorsalis* through out the activity season in Pakistan. Mango verieties Sonehra and Beganpalli were preferred over Sindhri, Desi, Almas, Langra in Pakistan; while number of *B. zonata* was more than those of *B. dorasalis*. (Qureshi *et al.,* 1991).

The population of *B. zonata* fluctuated considerably and peaked on 30th, 42nd, 45th and 8th and 13th standard weeks in Gujarat. It was positively correlated with maximum temperature. The peak was observed during May and July. (Sarada *et al.,* 2001, Babu and Viraktamath, 2003).

8.2.5. Identification

The adult of this species can be easily identified by the reddish brown colour of the thorax with a pair of parallel lateral vittae ending after extra alar setae. The abdomen is fuscous coloured with prominent T-shaped band. Male has pectin on third abdominal segment. The scutum is without medial vitta and basal setae. Small circular to oval shaped spots are found in antennal furrow. Wings are hyaline with a discontinuous costal band. This is differentiated by the presence of presutellar setae and transverse band on abdominal tergum, whereas a similar species, *B. affinis* lacks prescutellar setae and prominent T-shaped band. In *B. affinis*, there is a spurious band on the III-IV abdominal segment.

8.2.6. Description of Adult

Face fulvous with two medium sized oval black spots. Scutum red-brown with a pale fuscous pattern posteriorly with lateral yellow or orange postsutural stripes, scutellum entirely pale yellow coloured except for a narrow dark red-brown basal band and with two scutellar setae. Wing lacks a complete costal band (reduced to an isolated apical spot), cubital streak absent but with a very small pale fuscous tint in the cell cup. Legs fulvous with apices of femora red-brown and hind tibiae pale fuscous to fuscous, mid tibiae each with an apical black spur. On abdomen there is usually a pair of dark marks on tergum III and no medial dark line except on tergum V. Male abdomen with pecten. The males are attracted to methyl eugenol.

8.3. *BACTROCERA CORRECTA* (BEZZI)

It is commonly known as guava fruit fly. The old names for guava fruit fly were *Chaetodacus correctus* Bezzi, *Dacus bangaloriensis* Agarwal and Kapoor, *Dacus dutti* Kapoor, *Strumeta paratuberculatus* Phillip.

B. correcta is distributed to the whole of Indian subcontinent and Thailand but has also invaded places distant from Asia. Its distribution in India is peculiar. If unchecked it has the potential to become a major pest of citrus, peach and several kinds of tropical and subtropical fruits. It has been recorded in India on mango (*Mangifera indica*), guava (*Psidium guajava*), peach (*Prunus persicae*), bael (*Aegle marmelose*), citrus (*Citrus* sp), ber (*Zizyphus jujube*), *Caressa carandas*, *Eugenia michelii* and sandalwood berries. The adults are attracted to newly opened termitarium. In India it usually occurs in association with *B. zonata* and *B. dorsalis* and has similar hosts. In recent years it has increased its intensity in some parts of India, dominating the other two species on guava and mango. Its number starts increasing by late June or early July, when other fruit fly species are still low in number. In Tamil Nadu this species is

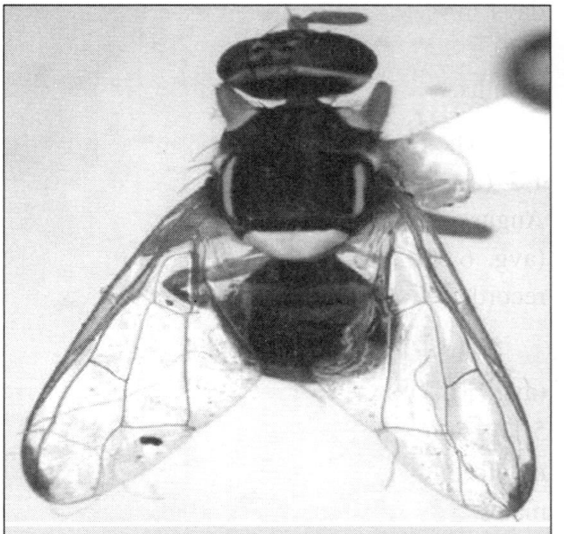

FIG. 30. *Bactrocera (B.) correcta* (Bezzi)

showing signs of dominancy over *B. Zonata* and *B. dorsalis* on guava for the first time, indicating that *B. correcta* may become a major threat to guava in south India (Kapoor, 2005).

8.3.1. Description of Adult

Bactrocera correcta is a small fly. It is recognized by a predominantly black scutum with lateral yellow stripes. Body a balanced mixture of black and yellow. Face fulvous with spot on transverse sulcus extended medially (linear) or with a transverse band. Hind tibia with keel like process on posterior-dorsal surface. Closer to apex, costal band discontinuous, breaking at cell $R_2 + R_3$ M. Pale pecten on third abdominal tergum. Two lateral pale whitish to yellow post sutural stripes present.

It is attracted to methyl eugenol.

Biology : Verma (2013) studied the biology of Guava fruit fly *B. correcta.* The salient features of the results on various aspects are as follows :

After mating egg laying started in 1-2 (average 1.50 ± 0.53) days in June-July, 2-3 days (average 2.60 ± 0.52) in August-September and egg laying ranged between 4-5 days (average 4.50 ± 0.53) in the month of October-November.

After oviposition, eggs hatched into 1st instar larva, the duration of them was recorded as 1-2 (avg. 1.60 ± 0.52), 2-3 (avg. 2.60 ± 0.52) and 3-4 (avg. 3.40 ± 0.53) days in the month of June-July, August-September and October-November, respectively. 2nd and 3rd instar were completed in 1-2 (avg. 1.50 ± 0.54) days and 2-3 (avg. 2.50 ± 0.54) days in month of June-July, respectively, 2-3 (avg. 2.50 ± 0.53) and 3-4 (avg. 3.40 ± 0.52) days in August-September and 3-4 (avg. 3.40 ± 0.48) and 4-5 (avg. 4.60 ± 0.52) days in October-November, respectively. Total length of maggot period was, 4-7 (avg. 5.30 ± 1.16), 7-10 (avg. 7.80 ± 1.48) and 10-13 (avg. 11.20 ± 1.23) days in respective months.

The pre pupal periods of guava fruit fly was noticed as 0.5-1 (avg. 0.75 ± 0.26) day, 1-2 (avg. 1.30 ± 0.48) and 2-3 (avg. 2.50 ± 0.53) days, in the month of June-July, August-September and October-November, respectively while the pupal periods of 6-8 (avg. 6.90 ± 0.80), 8-10 (avg. 8.80 ± 0.79) and 12-15 (avg. 13.30 ± 1.16) days were recorded in the respective months.

The longevity of adult of *B. correcta* was recorded as 5-19 (avg. 11.90 ± 4.89), 6-24 (avg. 15.40 ± 6.26), and 8-35 (avg. 20.00 ± 8.56) days in males and 6-20 (avg. 12.40 ± 5.17), 6-26 (avg. 13.20 ± 6.05) and 8-40 (avg. 22.30 ± 10.81) days in females during the month of June-November. The females of fruit fly lived for long time in comparison to males.

The total life cycle of guava fruit fly was completed in 15-18 (avg. 16.30 ± 1.16) days for males and 18-22 (avg. 19.50 ± 1.35) days for females in the month of June-July. In August-September total life cycle was completed in 18-22 (avg. 19.50 ± 1.35) and 20-24 (avg. 21.80 ± 1.62) days in males and females, respectively. Total life cycle was completed in 20-30 (avg. 24.10 ± 1.15) days (males) and 22-35 (avg.27.90 ± 4.65) days (females) in the month of October-November. The life cycle of *B. correcta* was completed rapidly at high temperature and decreased in low temperature at the all stages like eggs, maggots, pupa and adults.

8.4. *BACTROCERA CARAMBOLAE* DREW AND HANCOCK

It is reported from Aandman Island in India. Its primary hosts include *Averrohoa carambola*, *Syzygium samarrangense* and secondry hosts are guava and mango.

8.4.1. Description of the adult

This species is similar to *B. dorsalis* and *B. occipitalis* but differs from both in having narrower lateral dark patterns on tergite 3-5. A more pointed apex of aculeus and usually a large elongate dark spot more common in females, on the fore femur postero apically.

FIG. 31. *Bactrocera (B.) carambolae* Drew and Hancock

8.5. *BACTROCERA CARYEAE* (KAPOOR)

This species is currently widely distributed to southern India and Sri Lanka. It is confined to the southern part of the Indian subcontinent. It attacks guava and mango.

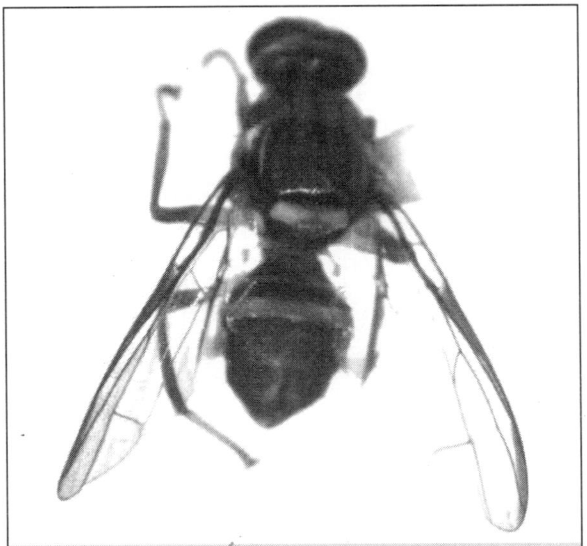

FIG. 32. *Bactrocera (B.) caryeae* (Kapoor)

8.5.1. Descriptionof the adult

Face yellow with black elongate oval spots. Scutum black, with lateral post sutural vittae tapering and ending before intra-alar. Legs fulvous with black spots on all femora and apex of femora darker. On wing a reduced band of microtrichia anterior to basal cell is found. Narrow costal wing band confluent with $R_{2+3,}$ anal streak narrow. It has very broad lateral longitudinal dark fuscous to black bands on abdominal tergite 3-5. Pecten of dark setae on tergum III of male present.

8.6. *BACTROCERA VISHNU* DREW AND HANCOCK

B. vishnu has been reported from Kodaikanal of Southern India. Its host is not known.

FIG. 33. *Bactrocera (B.) Vishnu* Drew and Hancock

8.6.1. Description of adult

It is similar to *B. carambolae* in the general colour pattern of the thorax, wing and abdomen. But it mainly differs in having a large elongate oval black spot on the apex of the forefemur.

8.7. *BACTROCERA LATIFRONS* (HENDEL)

B. latifrons is widely distributed in the oriental region. It is a serious pest of solanacious crops including chilly and bittergourd in Himachl Pradesh. In the Indian subcontinent it has been reared from egg plant (*Solanum melongena*) and *S. sisymbriifolium* (Narayanan and Batra 1960). It has been reared in berries of *Solanum vivarum* in Karnataka. It is predominant in southern India and it is a potential threat for other parts of India. It was introduced in Hawaii in 1983 (Vargas and Nishida, 1985).

FIG. 34. *Bactrocera (B.) latifrons*

8.7.1. Description of the adult

It is easily recognized by face with two dark spots. It has predominantly black scutum with two lateral postsutural yellow vittae. Scutellum entirely pale yellow with two scutellar setae. Wing with a complete costal band expanded into an apical spot at apex, cubital streak present. Abdomen oval, orange-brown to fuscous, tergum III with a basal transverse dark band and sometimes with a medial stripe on terga III-V. Male with pecten on tergum III. Its attractant is not known.

8.8. *BACTROCERA AFFINIS* (HARDY)

8.8.1. Description of adult

Head dark red-brown to fuscous, fulvous laterally and face shining black. The scutum is shining black without pale markings. Two narrow lateral yellow postsutural vittae, narrowing posteriorly to end well before infra alar setae are present. Scutellum yellow except for a moderately broad black basal band with two scutellar setae. Femora mostly shining black, mid femora entirely shining black, hind femora with basal 2/3 fulvous and apical 1/3 shining black, mid tibiae each with an apical black spur. Wings colourless except fuscous cell Sc, narrow fuscous costal band confluent with R_{2+3} and remaining narrow to end just beyond extremity of R_{4+5}. A narrow fuscous cubital streak is present. Abdomen oval, terga I and II are black except for a narrow red-brown band along inter-segmental line and a broad transverse fulvous to red brown band across posterior 1/2 of tergum II and narrowing to lateral margins; terga III-V red-brown except for a broad black band across anterior margin of tergum III. A pair of oval black shining spot is found on tergum. Male with pecten on tergum III.

Attractant : Cue lure.

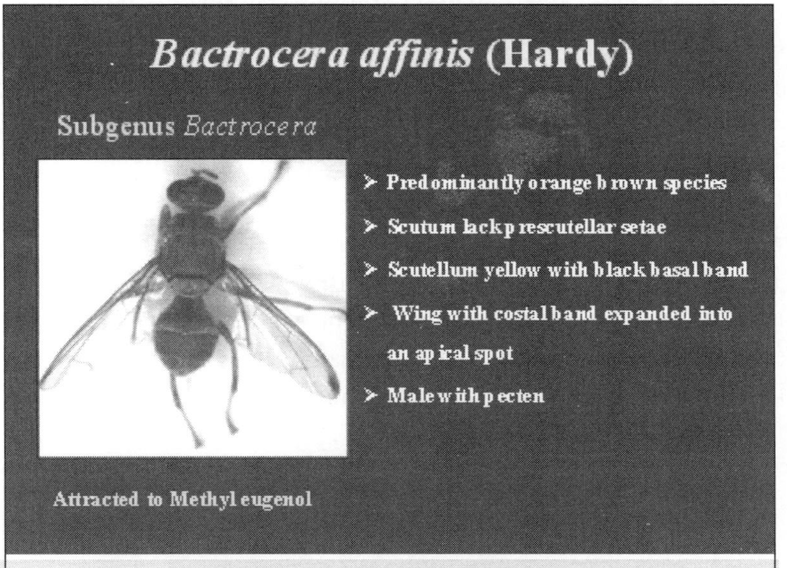

FIG. 35. *Bactrocera (B.) affinis*

8.9. *BACTROCERA NIGROTIBIALIS*

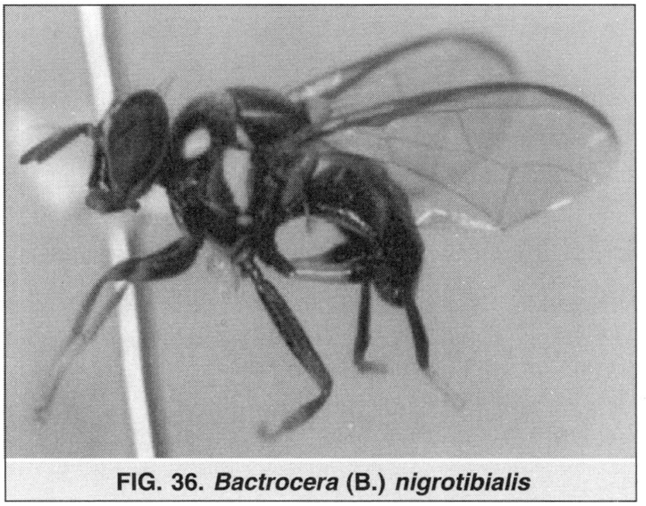

FIG. 36. *Bactrocera* (B.) *nigrotibialis*

8.9.1. Description of adult

Face shining black. Scutum black, lateral post sutural vittae ending well before intra-alars. Femora entirely black except basal half of hind femora. Abdomen black except for a yellow band at hind margin of tergum 2. Male pecten on third abdominal tergum. It is attracted to Cue lure.

8.10. *BACTROCERA NIGROFEMORALIS* (WHITE AND TSURUTA)

It is a new record from Himachal Pradesh as well as from north India.

8.10.1. Description of adult

It is a black coloured fruit fly. Face shining black. Scutum black, lateral post sutural vittae ending well before intra-alars. Femora entirely black except basal half of hind femora. Abdomen black except for a yellow band at hind margin of tergum 2. Male pecten on third abdominal tergum. It is attracted to cue lure.

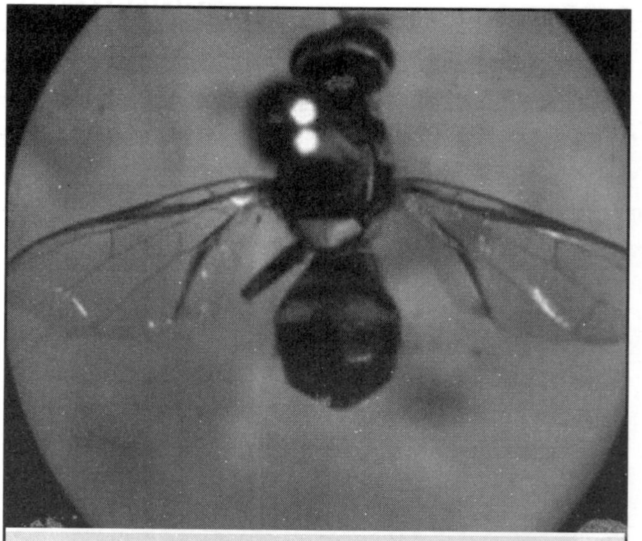

FIG. 37. *Bactrocera (B.) nigrofemoralis*

8.11. *BACTROCERA VERBASCIFOLIAE*

8.11.1. Description of adult

Circular black spots on fulvous face. Wings hyaline with costal band from cell Sc to wing apex beyond vein R_{4+5}, overlapping vein $R_{2+3.}$ Abdomen generally orange brown with a narrow medial longitudinal black band on terga 3-5 and black antero-lateral margins. Male pecten on third abdominal tergum. It is attracted to methyl eugenol.

FIG. 38. *Bactrocera (B.) verbascifoliae*

8.12. *BACTROCERA VERSICOLOR* (BEZZI)

8.12.1. Description of adult

Scutum predominantly orange brown, with slightly semi-circular lateral vittae that just reach or end before intra alars. Scutellum yellow with a black watery spot at apex of scutellum. Male pecten on third abdominal tergum. It is attracted to methyl eugenol.

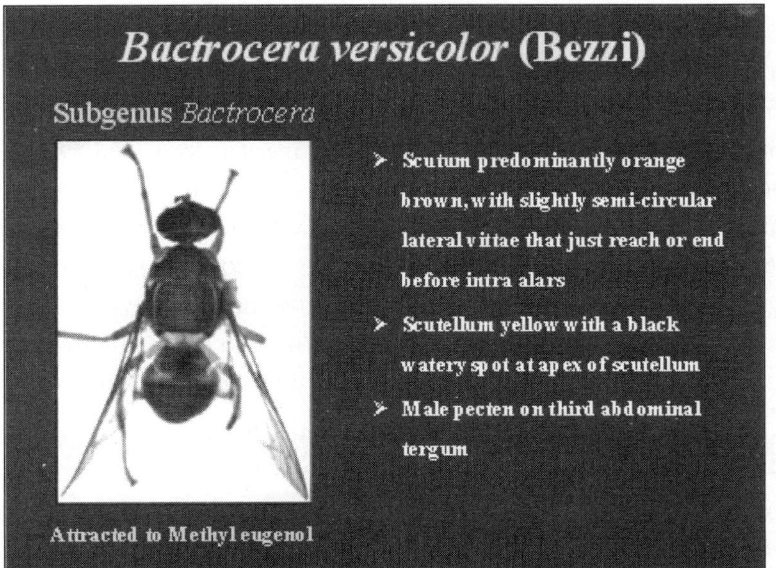

FIG. 39. *Bactrocera (B.) versicolor*

8.13. *BACTROCERA PARAVERBASCIFOLIAE* DREW AND RAGHU

It is recorded from south India. Its host is not known.

8.13.1. Description of adult

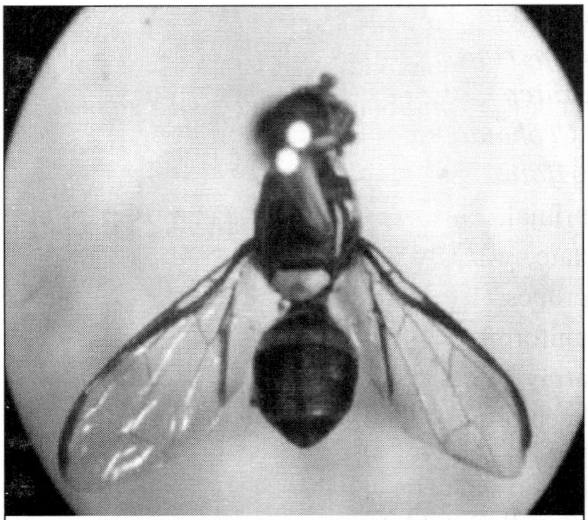

FIG. 40. ***Bactrocera (B) paraverbascifoliae***

Face fulvous with medium sized circular to oval black spots. Scutum black with red-brown to dark red-brown below and behind lateral postsutural vittae. Two parallel sided lateral postsutural vittae of medium width. Scutellum yellow except narrow black basal band with two scutellar setae. wings colourless except for dark fuscous cell sc with a narrow fuscous to dark fuscous costal band confluent with R_{2+3} and gradually widening to end between extremities of R_{4+5} and M and a narrow fuscous cubital streak. Legs are fulvous except a small area of dark fuscous on base of fore tibiae and fuscous on base of mid tibiae. Hind tibiae dark fuscous basally to fuscous apically. Mid tibiae each with an apical black spur. Abdomen oval, tergum I fuscous anterocentrally, tergum II red-brown except for a narrow transverse black anteriorly which ends laterally just before the dark fuscous lateral margins, terga III-V orange-brown except for a distinct black 'T' pattern. A pair of oval red-brown shining spots on tergum V. Male with pecten on tergum III.

Attractant : Methyl eugenol

Subgenus : ***Zeugodacus***

8.14. BACTROCERA (*ZEUGODACUS*) *CUCURBITAE* (COQUILLETT)

B. cucurbitae is co mmonly known as "Melon fruit fly" due to its preference for melons. In India it destroys either partially or completely, over 60% of cucurbit crops.

8.14.1. Distribution

It is distributed throughout India. Outside India, it has been recorded from Nepal, Pakistan, Sri Lanka, Myanmar, Malysia, Java, China, Philipines, Hawaiian Island, Australia and East Africa.

8.14.2. Host plants

It attacks mostly cucurbits but the fly has been recorded in India on pumpkin (*Cucurbit maxima* and *C. moscata*), cucumber (*Cucumis melo* var. *momordica*), long melon, (*Cucumis utilisimus*), wild cucurbit (*Cucumis trigonis*, and *C. pubescencs*), water melon (*Citrullus vulgaris* var. *fistulosus*), bottle gourd (*Langenaria vulgaris*), bitter gourd

(Momordica charantia), ribbed gourd (*Luffa aculangula*), sponge gourd (*Luffa cylindrica*), snake gourd (*Trichosanthes anguina*), wild snake gourd (*T. cumerina*), praval (*Trichosanthes dioica*), *Coccinia indica*, kundru (*Cephalandra indica*), galls of *Vitis trifolia*, tomato,watery rose apple, brinjal, chilli, papaya, guava, peach, date, citrus and wild vine with red drupes. It is the only species that is uniformly widespread, attacking a large array of cucurbit fruits in India. It has been observed to withstand all kinds of adverse climatic conditions from the plains to the hilly tracts.

FIG. 41. ***Bactrocera (Z.) cucurbitae*** **(Coqillett)**

It is reported to be associated with about 150 host plants (Kapoor, 2005). Batra (1968) recorded *B. cucurbitae* attacking cucurbit flowers and adults resisting in swarms in association with other *Bactrocera* species on the underside of guava leaves during winter. During the hot summer months (*i.e.*, April-June) attacks on cucurbits by B. cucurbitae often occur together with *Dacus* (Didacus) *ciliates* Loew. During July-August infestations of *B. cucurbitae* are so severe that almost all cucurbits are attacked resulting in heavy damage. In some cases, the entire crop is lost. The most vulnerable hosts are bitter gourd, ribbed gourd, bottle gourds, long melons, squash melons, snap melons, and cucumber (Agrawal and Kumar, 2008). The infestation continues until the appearance of fruit crop in October. Intensity of attack diminishes considerably after October, except sometimes in American chillies and tomatoes. In Banglore *B. cucurbitae* also damages musk melons from January to March (Sarode *et al.*, 1981).

8.14.3. Damage

The damage to crops caused by melon flies is due to :

1. Oviposition in fruit and soft tissues of vegetative parts of hosts.
2. Feeding by the larvae.
3. Decomposition of plant tissue by invading secondary microorganisms.
4. It has also been reported to develop in stem galls induced by gall midges (*Lasiosptera toombi* (Grover); Diptera : Cecidomyiidae on ivy gourd (*Coccinia grandis*) (Bhatia and Mahto, 1968).

The fly prefers green and tender fruits and is not able to pierce the hard rind of the larger fruits. Similarly, some of the gourds with hard rind escape damage. The fly sometimes

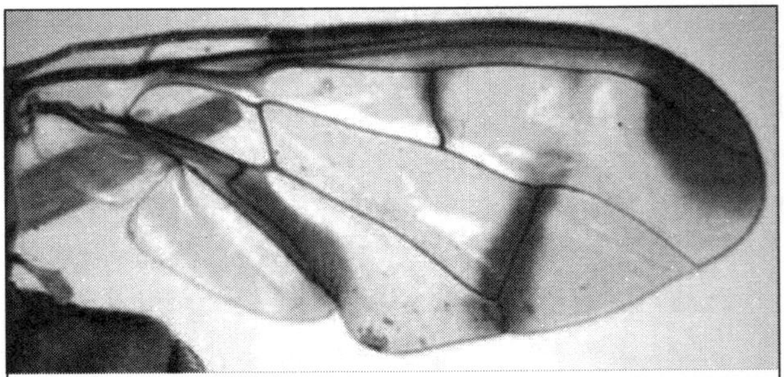

FIG. 41. Wing of *B. cucurbitae*

lays eggs in the corolla and other organs of the flowers and the maggots that hatch out of these feed on the flowers and a few on stems of cucurbit vines forming the galls. Larvae feeding inside fruits are most damaging. Mature fruits are attacked most which develop a water soaked appearance. Young fruits become distorted and usually drop. The larval tunnels provide entry points for bacteria and fungi that cause the fruit to rot. These maggots also attack young seedlings, succulent tap roots of watermelon, and stems and buds of host plants such as cucumber, squash and others.

8.14.4. Biology

Oviposition occurs about 10 days after emergence and continues at intervals. A female may lay about 300 eggs in natural conditions. Females prefer to oviposit in new plant growth such as young seedlings, growing tips and developing ovaries of all cucurbits except young cucumbers. Ripe fruits are preferred. Because of their high egg laying capacity and mobility, each female is capable of destroying large number of fruits in its life span. Adults generally live for 10 months to a year.

The life cycle from egg to adult requires 14-27 days. The eggs of the melon fly are slender, white and measure 0.2 cm in length. Eggs are inserted into fruit in bunches of 1 to 37. They hatch in 2 to 4 days. There are 3 larval stages, which are cylindrical, elongate, narrowed and somewhat curved downward at the end and mouth hooks at the head. These maggots reach approximately 1.25 in lenghth at maturity. The larval period lasts from 6-11 days. Pupation takes place in the soil beneath the host plant. They are elliptical and dull white to yellowish brown in colour. They may have ringed yellow bands around each segment. Pupal stage lasts for 9-11 days in summer.

8.14.5. Description of adult

Adult flies are slightly larger than house flies. The head and eyes are dark brown. The body yellowish brown with a yellow spot above the base of of the first pair of legs. A yellow stripe with curved line on either side is present down the centre of the back. The brown band extending along the leading edge, ending in a larger brown spot at the tip. A brown spot occurs near the wing margin. Abdomen is redish yellow with darker bands on the second and third abdominal segments. Legs are yellowish in colour.

Face fulvous with a pair of medium sized elongate oval black spots. Scutum entirely red-brown or fuscous to dark fuscous markings with both lateral and medial yellow stripes. Scutellum yellow with a narrow dark fuscous basal band, one pair scutellar setae and rarely two pairs. Wing with complete costal band, which is expanded into a spot at apex, crossvein dm-cu and r-m covered by infuscate area, broad fuscous cubital streak present. Legs with femora fulvous except for apical dark patterns which are red-brown on fore femora and fuscous to dark fuscous on mid and hind femora, fore tibiae fuscous, mid tibiae fulvous with fuscous basally while hind tibiae dark fuscous. All tarsi are fulvous. On abdomen a transverse band across tergum III is found and a medial longitudinal stripe is present on terga III-V. Males with a row of setae (the pecten) on tergum III. Males are attracted to Cue lure.

8.15. *BACTROCERA CAUDETA* (FABRICIOUS)

This species has also been known as *Bactrocera maculipennis* Doleschall, *Chaetodacus caudatus* Fabricious and *Zeugodacus caudatus* (Fabricious). It is distributed in Indonesia (Java, Sumatra), Malaysia, Myanmar, Taiwan, Thailand, Vietnam and Brunei. The species was also recorded fron India and Nepal.

8.15.1. Description of adult

It is easily recognised by being predominentle black with lateraland medial yellow scutal stripes. Face fulvous with a black line across mouth opening or with a black spot in the antennal furrows extended laterally and almost forming a line across the mouth opening. Scutum black with yellow medial post sutural vittae narrowed anteriorly and broadly triangular posteriorly. Wings hyaline, with costal band confluent with R_{2+3} expanded apically with an apical spot. Male pecten on third abdominal tergum. Its males are attracted to cue lure.

FIG. 42. *Bactrocera (Z.) caudata* (Fabricious)

The species is similar to *B. tau* but later differs in predominantly orange brown scutum and costal band extended into a comparatively larger apical spot and the two oval spots clearly separated.

8.16. *BACTROCERA TAU* (WALKER)

B. tau is very similar to *B. caudata*. It has been known as *Chaetodacus tau* (Walker), *Dacus caudatus* var. *nubilus* Hendel, *D. hagani* Meijere, *D. nubbilus*, Handel, *D. tau* (Walker), *Dasynura tau* Walker, *Zeugodacus bezzianus* Hering and *Z. nubilus*. It is widely distributed to the oriental region. It attacks most of the cucurbits, tomato, flashy fruits such as jack fruit, guava, mango, sapodilla and wax apple. The species is distributed to whole of the Indian subcontinent mostly in hilly tracts of northern India.

FIG. 43. *Bactrocera (Z.) tau* (Walker)

8.16.1. Description of the adult

The species is easily recognized by the orange brown scutum marked with lateral and medial yellow vittae present. Face fulvous with medium sized black oval spots,

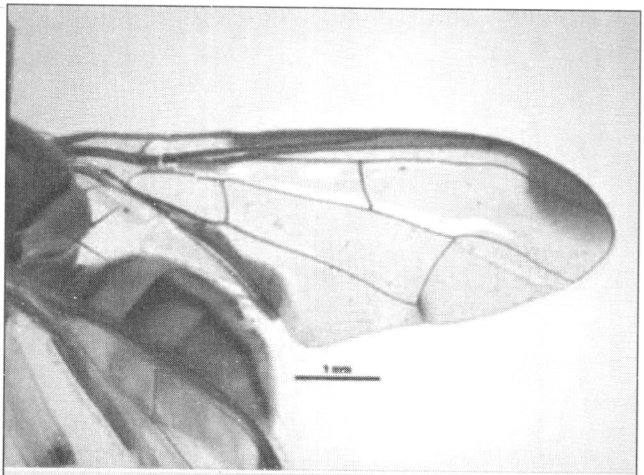

FIG. 44. *Wing of B.tau*

which are not interconnected. Scutellum entirely yellow with four setae. Wings with a narrow dark fuscous complete costal band overlapping vein R_{2+3} and expanding into a distinct apical spot at wing apex, cubital streak dark and broad. Abdominal terga III-V fulvous with a black T pattern. Male with pecten on tergum III. Males are attracted to Cue lure.

8.17. *BACTROCERA SCUTELLARIS* (BEZZI)

It is reported on all cucurbits especially on *Cucumis sativus* and *Cucurbita maxima.* It is distributed to whole of Indian subcontinent, Myanmar and Thailand. The species is active in cool climates and is becoming an important fruit fly pest in some areas of Himanchal Pradesh.

FIG. 45. *Bactrocera (Z.) scutellaris* (Bezzi)

8.17.1. Description of adult

Face fulvous with a pair of transverse oval black spots pointed towards centre. Scutum shining black with narrow lateral and median postsutural yellow vittae. Scutellum yellow with a distinct black apical spot and two pairs of scutellar setae. Wings with a narrow dark fuscous complete costal band which is extremely narrow beyond apex of vein R_{2+3} and expanding into a distinct apical spot around apex of R_{4+5}, cubital streak dark and broad. Abdominal terga III-V mostly dark fuscous to black. Male with pecten on tergum III. The female is provided with teeth on each denticles of the eversible membrane of ovipositor which are comparatively narrow with rounded apices. It is attracted to Cue lure.

8.18. *BACTROCERA OLEAE* GMELIN

Fletcher (1917) first reported it from India and Pakistan. Later, Pruthi and Bhatia (1938) and Narayanan and Batra (1960) reported *B. oleae* from North West India on wild cultivated olives. It is distributed to most of Europe, Africa, Middle East, northern India and north-West Pakistan. It is a serious pest of olives and involves much money to control it. According to Kapoor (2005), it is the single species included in Subgenus *Daculus* while some include it in Subgenus *Zeugodacus*.

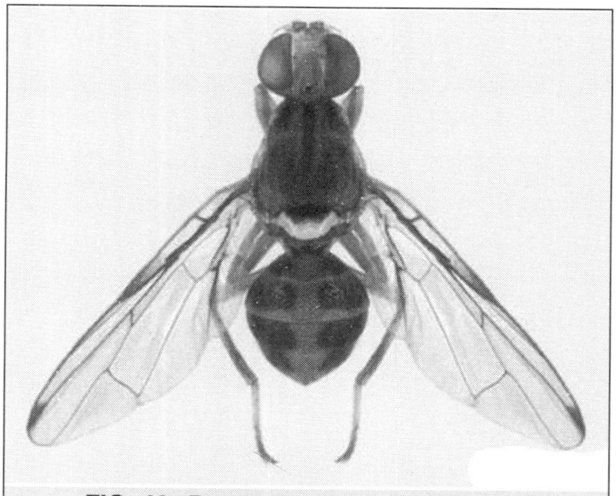

FIG. 46. *Bactrocera (Z.) oleae* Gmelin

8.18.1. Description of adult

It can be easity recognized by the predominant black scutum (sometimes this colour does not extend laterally, making the lateral area red brown, especially in African specimens. Facial spots present. Thorax is without yellow lines; abdomen orange medially and black laterally. Wings with reduced pattern.

FIG. 47. *Bactrocera (Z.) gavisa*

8.19. *BACTROCERA GAVISA* (MUNRO)

8.19.1. Description of adult

Face fulvous with black spots. Lateral vittae extending beyond intra alars to touch scutellum. Mesopleural stripe inverted L-shaped. It is attracted to cue lure.

Subgenus : *Hemigymnodacus/Paratridacus* Shiraki.

8.20. *BACTROCERA (HEMIGYMNODACUS) DIVERSA* (COQUILLETT)

It is a major species widely distributed in India, Thailand, Malaysia, Sri Lanka, Egypt, Taiwan and Brazil. This fly was originally known in South India and Sri Lanka but now has invaded many other locations in India, Pakistan and Thailand. On the whole the damage due to this fly in this part of world is low due to the dominance of *Bactrocera cucurbitae*. It attacks a variety of fruits and vegetables such as guava, mango, white gourd, tori, ivy gourd, bottle gourd and pumpkin and various gourds. It also attacks sour orange, jamun. According to Batra (1953) this fly breeds in the flowers of gourd and *Luffa* sp. Sometimes banana and citrus are also attacked. Its presence in the field coincides with the *B. cucurbitae* and *B. zonata*, which restricts its increase in number. It usually congregates on the underside of leaves of guava and mango along with *B. dorsalis* and *B. zonata*.

8.20.1. Damage

The dark puncture caused by the oviposition of adult fly is not very conspicuous as its colour blends with the colour of fruit. The maggots on hatching feed on the pulp of the

FIG. 48. *Bactrocera (H.) diversa (Coquillet)*

fruit for few days and a brown rotten patch appears on fruit surface. The Infested fruit finally fall on the ground.

8.20.2. Biology

This fly overwinters in the adult stage. The flies are active in spring. They breed on the flower of cucurbitaceous plants and then migrate to orchards. In the rainy months the optimum temperature and humidity enables them to compete the life cycle quickly and there is a steep rise in population. The eggs hatch in 1to 4 days, the larval period lasts for 4 to 5 days and pupal period may continue for 7 to 13 days.

8.20.3. Description of adult

The adult flies have greenish black thorax with 3 longitudinal, lemon yellow stripes. The costal and anal bands are narrow and weekly developed.

Face of male entirely pale yellow without black spots while face of female with a transverse black line above mouth. Scutum black with broad parallel sided lateral yellow vittae and a narrow medial postsutural vittae. The scutellum is yellow with a narrow black basal band and 2 or 4 scutellar setae. Wing with a narrow dark costal band confluent with R_{2+3} and widening across apex of wing, a dark cubital streak present. Abdominal terga III-V red-brown with a black T pattern. Male without pecten. It is attracted to methyleugenol.

Subgenus : *Javadacus* Hardy.

8.21. *BACTROCERA (JAVADACUS) TRILINEATA* (HARDY)

The host is not recorded so far.

8.21.1. Description of adult

Face entirely fulvous without black marking. Scutum black with broad parallel sided lateral yellow vittae and a narrow medial postsutural vittae, anterior supra alar seate absent. Scutellum yellow with a narrow black basal band and 2 scutellar setae. Wings with complete costal band not overlapping R_{4+5} and expanding slightly at wing apex. Legs fulvous with apices of femora red-brown. Abdominal terga I and II black except for a narrow red-brown band along intersegmental line and a broad transverse fulvous to red brown band across posterior 1/2 of tergum II and narrowing to lateral margins. Terga III-IV entirely black except fulvous at postero-submedially part of terga IV, terga V entirely fuscous to light brown. Attracted to methyl eugenol.

FIG. 49. *Bactrocera (Javadacus) trilineata* Hardey

(B) GENUS DACUS

Subgenus : ***Didacus***

8.22. *DACUS (DIDACUS) CILIATUS* LOEW

This species is commonly known as Ethiopian fruit fly. In Indian subcontinent it has been known as *Dacus brevistylus*. It prefers melons therefore, also known as Ethiopian melon fly. It was first recorded from some parts of India by Bhatia (1939), infesting large variety of melons and wild cucurbits. Its most preferred hosts are squash melon and ivy gourd. The species was in low profile due to the dominancy of *B. cucurbitae*. It is found in some parts of India, Nepal and Pakistan. The increasing attack on squash melon, bitter gourd and cucurbits in some parts of northern India is of much concern to vegetable growers.

8.22.1. Description of adult

The species can be easily recognized in being predominently orange with face fulvous with two black spots. Scutum reddish brown without lateral or medial vittae. Scutum without anterior supra alar setae and pre scutellar setae. Wings with costal band confluent to R_{2+3} but dilated apically near and beyond vein $R_{4+5.}$ Abdominal tergites fused to form hemispherical capsule; with oval black spots on either side of tergite 3. Males usually with two round black spots on abdominal terga. The females have numerous teeth in each denticle on the eversible membrane of the ovipositor which are closely set and comparatively slender. These are not attracted to methyl eugenol or cue lure.

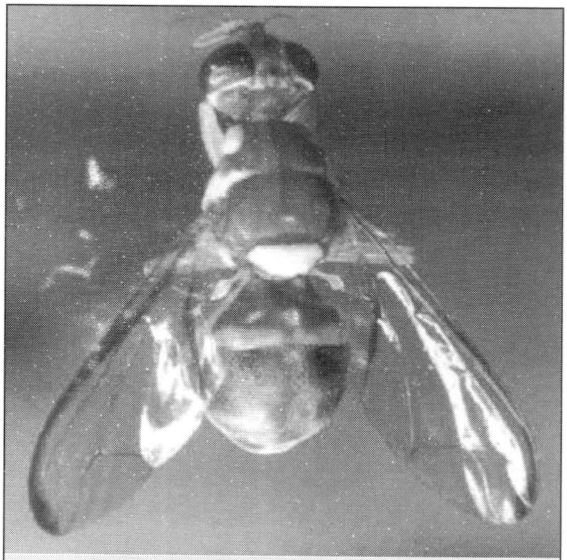

FIG. 50. *Dacus (Didacus) ciliatus* Loew

Subgenus : *Callantra*

8.23. *DACUS (CALLANTRA) RAMANI*

Host is not known.

FIG. 51. *Dacus (Callantra) ramanii*

8.23.1. Description of adult

Face pale brown with a transverse black line above mouth. Scutum red brown with a narrow black line at centre. Scutellum yellow without a narrow black basal band and two scutellar setae. Wings with a pale fuscous tint across membrane except for fuscous cell Sc, broad fuscous costal band confluent with R_{4+5} and expanding apically into a large round dark spot which overlaps M, pale fuscous anal streak present. Legs with fore femora red-brown, mid and hind femora fulvous except red-brown around apical 1/3; fore, mid and hind tibiae red-brown. Abdomen elongate-oval and petiolate. Abdominal terga I-II pale brown with black or dark brown at middle, terga III-V red-brown except for a narrow transverse black band across anterior margin of tergum III. It is attracted to cue lure.

8.24. *DACUS (CALLANTRA) LONGICORNIS* WIEDEMANN

The host is not recorded for this fly.

8.24.1. Description of adult

Face fulvous with a pair of oval black spots. Scutum dark red-brown without distinct black patterns, lateral and medial postsutural vittae absent. Scutellum yellow with broad red-brown basal band and two scutellar setae. Wing with cells bc and c fuscous, a broad dark fuscous complete costal band overlapping vein R_{4+5} for its entire length. Legs with fore femora dark red-brown to fuscous, mid femora dark red-brown to fuscous

FIG. 52. *Dacus (Callantra) longicornis* Weidemann

except basal ¼ fulvous, hind femora dark fuscous, fore and mid tibiae dark red–brown to fuscous, hind tibiae dark fuscous. Abdomen elongate-oval, strongly petiolate, abdominal terga III-V dark fuscous to black, large orange brown spots posterocentrally on terga IV and V with oval shining spots on tergum V. Male with pecten on tergum III. Males are attracted to Cue lure.

8.25. *DACUS (CALLANTRA) SPHAEROIDALIS* (BEZZI)

Host of this species is not recorded.

FIG. 53. *Dacus (Callantra) sphaeroidalis* (Bezzi)

8.25.1. Description of the adult

Face red brown with two large elongated black spots. Scutum red brown without dark markings. Scutellum yellow with a narrow black basal band and 2 scutellar setae. Wings with a pale fuscous tint across membrane except for fuscous cell sc, broad fuscous costal band confluent with R_{4+5} and expanding apically in to a large round dark spot which overlaps M; pale fuscous anal streak present. Legs with fore and mid femora red-brown, hind femora fulvous except red-brown around apical 1/3, fore and mid tibiae red-brown, hind tibiae red-brown basally. Abdomen elongate-oval, petiolate but not as strongly as in *D. longicornis*. Abdominal terga III-V red-brown except for a narrow transverse black band across anterior margin of tergum III, a pair of oval fuscous to dark fuscous shining spots on tergum V. Male with pecten on tergum III. Attracted to Cue lure

Subfamily : TEPHRITINAE

Tribe : GASTROZONINI

(C) GENUS : *CYRTOSTOLA* HANCOCK AND DREW

8.26. *CYRTOSTOLA LIMBATA* (HENDEL)

The host is unknown.

FIG. 54. *Cyrtostola limbata* (Hendel)

8.26.1. Description of adult

Scutum with four longitudinal black vittae placed submedially and sublaterally. Scutellum with a large apical black spot with four scutellar setae. Wing with fulvous basal area, including most of cell sc (which is brown basally) and narrow transverse and marginal brown bands, across r-m crossvein reaching costa well distal of cell sc, that across dm-cu cross vein converging with and weakly joined to the subapical band over apex of vein M. Abdomen with transverse black bands anteriorly on terga III-V.

Subfamily : TEPHRITINAE

Tribe : PLIOMELAENINI

(D) GENUS : *PLIOMELAENA* BEZZI

8.27. *PLIOMELAENA UDHAMPURENSIS* (AGARWAL AND KAPOOR)

It is recorded on cucurbits.

8.27.1. Description of adult

It is a small sized, blackish-brown species. Head with width more than height. Thorax black with yellow dust and small yellow pubescence. Scutellum wider than long, rounded at apical end, black, with yellow dust, lateral margins of scutellum yellow. Four scutellar bristles, all equal in size. Wings hyaline at base, marking dark brown, costal spine two, both cross veins straight, wing marking light in axillary lobe. Legs yellow, fore tibiae with long yellowish-white bristles. Abdomen black-brown with few small yellow pubescence on abdomen. Attractant is unknown.

Tribe : TEPHRITINI

FIG. 55. *Pliomelaena udhanpurensis* (Agarwal & Kapoor)

(E) GENUS : *DIOXYNA* FREY

8.28. *DIOXYNA SORORCULA* (WIEDEMANN)

It is recorded on cucurbits.

FIG. 56. *Dioxyna sororcula* (Weidemann)

8.28.1. Description of adult

It is a small species with head longer than height. Body black in ground colour and covered with dense yellow-grey dust and scale-like setae. Two pairs of frontal and orbital setae; proboscis slender, elongate and geniculate. Two scutellar setae. Wing markings as 3 large hyaline spots filling cell r1 and 4-5 hyaline round spots in cell Rm. Its attractant is unknown.

8.29. *CARPOMYIA VESUVIANA* COSTA

It is known as Ber fruit fly which is primarily a Palaearctic species widespread in Indian subcontinent. It is specific to Indian jujube (*Zizyphus mauritiana)*, common jujube (*Z. jujube*) and *Z. nummularia*. It also attacks wild bers such as *Z. rotundifolius* and *Z. sativus*. The fly causes considerable damage to jujube, sometimes 90% in northern India.

FIG. 57. *Carpomyia vesuviana* (Ber fruit fly)

8.29.1. Description of adult

The fly is easily distinguished in having distinctive patterns of black marks on the scutum, and distinct yellow and brown bands on the wings.

❑❑❑

9

Chapter

Management of Fruit Flies

Present management stratagies mostly focus on chemical insecticides. However, the treatments are costly and may affect the quality of the products. Also restrictions or bans on certain chemical treatments have made the certification of commodities for export difficult. In order to minimize the indiscriminate and injudicious use of chemical pesticides, integrated pest management (IPM) has been enshrined as cardinal principle of plant protection in the overall crop protection programmes. IPM is an eco-friendly approach for managing pest problems encompassing available methods and techniques of pest control such as cultural, mechanical, biological and chemical in a compatible and scientific manner. The greater emphasis has been given on biological control including use of bio-pesticides. IPM is also useful in reducing the pesticides residues in exportable agricultural commodities. IPM package of practices for mango and guava crops will be helpful to minimize the ill effects of chemical pesticides to promote the IPM for sustainable production. These packages will be useful for the researchers, extension workers and farmers alike who are engaged in the agricultural practices.

9.1. FOOD BAIT

Food baits can be prepared to attract different fruit flies. Average number of fruit flies caught per trap by using four traps with food baits between May and August 1997 were *B. dorsalis* and *B. zonata* (Kumar and Agarwal, 1998).

TABLE 6 : Different food baits and species caught

Clove oil (2 ml) + Malathion 50 EC (1 ml) mixed with	*dorsalis*	*zonata*
Protein hydrolysate (20 gm)	44	148
Ripe mango pulp (20 gm)	28	132
Fermented palm juice (20 ml)	9	76
Sugar (20 gm)	10	16

The attractiveness of melon fly *Bactrocera cucurbitae* (Coq.) to soybean hydrolysate and banana were 85-95% more as compared to fish meal, beef extract, bread and dog buiscuit. Grape and banana pulp was more attractive when beer and palm oil were mixed in the bait (Bharathi *et al.,* 2004).

The MLT (Multilure traps) with yellow base, baited with hydrolysed protein attracted higher number of *B. oleae* males, while Torula yeast attracted higher number of female flies in Spain.

Food bait mixtures, are often more attractive than a single compound in attracting insects and flies respond positively to the sources of ethyl alcohol and acetic acid. Food attractants, even the non standard, were evaluated in investigations carried by Kumar (2018) as alternatives to synthetic pheromones and parapheromones to lure and trap fruit flies. This research has indicated that protein hydrolysate and casein are suitable attractants, especially when mixed with ammonium acetate and pulps of guava and mango which are cheaper and easier to make base of food baits. Bait admixtures, containing protein hydrolysate, guava or mango pulp and sugar in equal ratio (1 : 1 : 1) with ammonium acetate was superior bait in attracting fruit flies. However, ammonium acetate as base

FIG. 58. Trap with food bait

without any protein was also proved as best bait to lure fruit flies. Casein protein baits with or without ammonium acetate was moderately attractive to fruit flies. The attractiveness of such baits lies in their microbial fermentation products which emit complex odour containing olfactory and phagostimulatory semiochemicals attractive to fruit flies. Since guava and mango contained mainly D-glucose, fruit flies prefer them to other sugars. However, as the food attractants are not as species specific as pheromone, other Dipteran flies belonging to other families are likely to get attracted to these baits in traps.

FIG. 59. Fruit flies trapped in food bait

In the present investigation, methyl eugenol was the most effective attractant which attracted five *Bactrocera* spp. *viz.*, *B. zonata, B. dorsalis, B. correcta, B. diversa* and *B. caryeae*.

9.2. ATTRACTANTS

Methyl eugenol traps outperform other protein baits as well as ammonium salts. But due to its high cost and being unapproachable to farmers, the baits were formulated by Kumar (2018). Effectiveness of protein hydrolysate + sugar + fruit pulp (1 : 1 : 1) along with 5% ammonium acetate was most promising in guava and mango, however, its effectiveness was almost half that of methyl eugenol. All attractants tested captured more number of males and females except traps with sugar and fruit pulp only. Future research is suggested in the development of stronger and female specific attractants for fruit flies.

Some of the most common chemical attractants are :

(i) Methyl Eugenol : 4 allyl-1, 2-dimethoxybenzene-carboxylate (ether) is highly attractive kairomone lure and is found in more than 20 plants belonging to various families, for example leaves of *Tulsi* and *Citronella* plants. This is the most powerful male lure and quite long lasting under suitable conditions. It attracts a large number of males of important Decine spp. like *dorsalis, zonata, correcta, diversa* and *cucurbitae.* In general three isotopes of methyl eugenol are used, 3-4 dimethoxy methoxymethyl benzene, 4 dimethoxy benzene and 3-4-dimethoxy methoxy methyl benzene.

FIG. 60. Methyl eugenol charged traps in guava and mango

(ii) Cue-lure : 4 (p-acetoxyphenyl)-2-butanone : it does not occur in nature. It is extremely good as male lure for *B. cucurbitae.* one of its analogue, Rasberry Ketone, a natural plant product, known as Willisons lure in Australia used for *B. tryoni* males, has also been found attracting *B. cucurbitae* in Hawaii but cue-lure has been the best.

(iii) Trimed-Lure : t-buty 1-4 (or 5)-chloro-2-methyl cyrboxylate : It is the best male lure for Mediterranean fruit fly, *Ceratitis capitata.*

(iv) Bacu-lure : It is a parapheromone which attracts adult of melon fruit fly (*B. cucurbitae*).

The above mentioned lures can be obtained from the following places :

Name of the lure	*Places*
Methyl Eugenol	1. SISCO Research Laboratories Pvt. Ltd., 2F, Satyam Industrial Estate 'C' Wing, Dr. Cardinal Grocious, Mumbai-400099 2. Spectrochem Pvt. Ltd., Mumbai 3. Fritzche, Dodge and Olcott Inc. 76, Ninth Avenue, New York, N.Y. 10011, U.S.A. 4. International pheromones system Ltd. Units 12 and 13, Meadows Lane, Industrial Estate Ellesmere Port. South Wirral L65 4 EH, England.
Cue-lure	Food Industries, Bromborough Port, Wirral, Cheshine L6L 45U, England.
Bacu-lure	Pest Control (India) Pvt. Ltd. Jagdamba House, Anupam Theatre, P.B. 9060, Goregaon (E) Mumbai.
Trimed-lure	UOP Chemical Division State Highway 17, East Rutherford, New Jersey, 07073, U.S.A.

9.3. BAIT SPRAY

The adult fruit can also be controlled by bait sprays of carbaryl (0.2%) + Protein hydrolysate (0.1%) or molasses starting at pre-oviposition stage (first week of April), and repeated once after 21 days.

9.4. PLANT EXTRACTS AS BIO-PESTICIDES

The intense application of synthetic chemical pesticides in different crops has created many complications such as pesticide resistance in insects, secondary pest out breaks and health hazards. Many alternatives are being proposed as eco-friendly and economically viable non chemical method, biopesticide, biological control, botanical control, culture control or traditional pest management. Ignacimuthu and Gabrial (2009) suggested that in the present scenario, there is a need for botanical pesticides to reduce the environmental pollution and health hazards created by synthetic chemical pesticides.

Verma (2013) reported effect of various indigenous plant extracts for the management of Guava fruit fly *B. correcta*. He reported that 8 per cent of Neem leaf extract was found highly effective with minimum emergence of adults from treated pupae. Only 2.22 per cent, 6.67 per cent and 8.89 per cent adults emerged at 6, 8 and 10 days of intervals, respectively. As regard mean of adult emergence is concerned, it was lowest in neem 8% (5.93 per cent) while Oak extract 8% registered maximum adult emergence of 36.30 per cent. However, all the treatments were found superior over control.

At 8% concentration of neem leaf extract (NLE), logavity of emerged adult was less as compared with other plant extracts, with minimum 11.11 per cent survival after 10 days, 6.67 per cent, at 15 days and 4.44 per cent at 20 days. The mean survival of 7.41

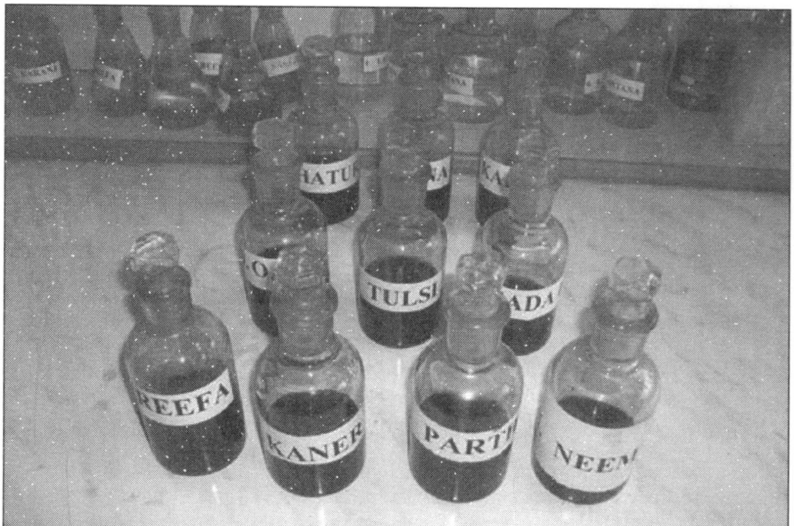

FIG. 61. Different indigenous plant extracts

FIG. 62. Guava fruits infested with Guava fruit fly *B. correcta*

FIG. 63. Larvae (Maggots) of Guava fruit fly *B. correcta*

per cent was recorded by neem leaf extracts followed by 11.11 per cent in Parthenium and 13.33 per cent in Dhatura extract. Oak extract was least effective with maximum 44.44 per cent mean survival of adults at 8 per cent.

The egg fertility was recorded after 4, 6 and 8 days of treatment of guava fruit, by fruit treatment against *Bactrocera correcta* (Bezzi) adults which was significantly lower in neem leaf extract fruits where 8.89, 6.67 and 4.44 per cent egg fertility was recorded after 4, 6 and 8 day of release of adults, respectively, at 8% concentration. The fecundity

FIG. 63. Laboratory experiments with guava fruit fly *B. correcta*

FIG. 64. Guava fruits treated with sandwich method

of guava fruit fly had been recorded maximum after 4 days of release which gradually decreased after 6 and 8 days.

Different instars of larvae of *B. correcta* were fed on treated slices of guava by sandwich method, at 2, 5 and 8 per cent concentration of indigenous plant extracts and mortality of maggots was recorded after 6, 8 and 10 days intervals. It was observed that 1st, 2nd and 3rd instar larvae registered maximum mortality by Neem (39.75 per cent) followed by Parathenium (36.79 per cent), Dhatura (33.09 per cent) and Karanj (30.37 per cent) at 5% concentration. However, at 8% concentration again Neem leaf extract

FIG. 65. Plant extracts as reproduction inhibitor

imparted maximum mortality of larvae with 31.11, 37.78, 73.33 per cent mortality for 1st instar, 28.89, 35.56, 71.11 per cent mortality for 2nd instar and 26.67, 33.33 and 53.33 per cent mortality for 3rd instar larvae after 6, 8 and 10 days of treatment, respectively.

1st, 2nd and 3rd instar at 8 per cent solution of neem leaf extract caused maximum larval mortality (43.46 per cent) while Oak and lantana were found least effective causing minimum mortality of 16.05 and 19.01 per cent. Among all indigenous plant extracts, neem formulation was found most effective. 1st instar was found more susceptible in comparison to 2nd and 3rd instar. This means that early stage of maggots can be managed by indigenous plant extracts.

9.5. BIOLOGICAL CONTROL

To the extent possible, bio-control agents should be carefully considered for their possible impact on non target organisms. Some general rules are as follows :

A. Predators are usually more generalised than parasites.

B. For biological agents host specification, including information on behaviour of related pest species is required.

C. Selection of related agents should be known, whether it be host specific or attack only closely related non-target species.

Information on the available biological control organisms are given in table form to allow comparisons between different Tephritid species as decision-making tool, and to help in the selection of the best combination of useful biological control agents.

TABLE 7 : List of common parasitoids and predators of fruit flies

Sl. No.	Parasitoid	Family	Hosts
1.	*Biosteres longicaudatus* (Ashmead)		*B. zonata*
2.	*Dirhinus giffardii* Silvertri		*B. zonata, B. dorsalis*
3.	*Bracon longicaudatus*	Braconidae	*B. dorsalis*
4.	*Spalangia grotiusi* Girault		*B. dorsalis*
5.	*Opius* Spp.		*B. dorsalis*
6.	*Diachasmimorphra longicaudata*	Braconidae	*B. zonata, B. dorsalis, D.ciliatus, C. visuviana*
7.	*Psyttalia fletcheri* (Silv.)	Braconidae	*B. Cucurbitae*
8.	*P. incise* (Silv.)	Braconidae	*B. Cucurbitae*
9.	*P. manii* (Silv.)	Braconidae	*D. ciliatus*

S.N.	Predators	Stage	Hosts
1.	Big headed ants (*Pheidole megacephale*)	Larvae	*C. capitate*
2.	Carbonine wasp (*Ectemnius timidoventris*)	Adults	*C. capitate*
3.	Anthocorid bug (*Orius insidiosus*)	Eggs	*B. dosalis*
4.	Agentine ants (*Iridomyrmex humilis*)	Larvae	*C. capitata*
5.	Fire ants (*Solenopsis geminata*)	Adults	*B. dosalis*

TABLE 8 : Natural Enemies of True Fruit Flies

Species	Parasites/Predators	Notes
	Opius persulcatus	Braconid parasite from India
	Opius oophilus	Braconid parasite from India
	Cothonaspis sp.	Braconid parasite from India
	Pseudeucoila	Braconid parasite from India
	Trybliographa daci	Braconid parasite from India
	Dirhionusgiffardii	Braconid parasite from India
	Spalangia spp.	Pteromalid parasite from India
	Syntomosphyrum indicum	Europhid parasite from India
	Trichopria sp.	Diapriid parasite from India
Batcrocera caudata		
Batcrocera ciliates	*Dirhionusgiffardii*	Chalcid pupal parasite from Pakistan
	Psyttalia incise	Braconid egg-larval parasite from India
	Galesus sp.	Diapriid parasite from India
	Pachycerepoideusvinde miae (= dubius)	Pteromalid parasite from India
	Spalangia endius (= Philippinensis)	Spalangiid parasite from India
Batcrocera correcta	*Dirhinu* ssp.	Chalcid larval-pupal parasite from India
	Diachasmimorpha longicadata	Braconid parasite from Thailand
	Fopius arisanus	Braconid parasite from Thailand
	Fopius persulcatus	Braconid parasite from Thailand
	Fopius vandenboschi	Braconid parasite from Thailand
	Psyttalia fletchri	Braconid parasite from Thailand
	Psyttalia incise	Braconid parasite from Thailand

	Psyttalia makii	Braconid parasite from Malaysia and Thailand
	Psyttalia sp. nr *fletchri*	Braconid parasite from Thailand
	Psyttalia sp. nr *makii*	Braconid parasite from Thailand
	Utetes bianchii	Braconid parasite from Malaysia and Thailand
Bactrocera cucumis		
Bactrocera cucurbitae	*Psytallia fletchri = opius fletchri*	Common Indian egg-larval Braconid parasite, in Hawaii restricted to Melon Fly. Also in the Philippines, Guiana, India and Ryukyu.
	Opius watersi	Egg-larval Braconid parasite in North India and Ceylon
	Opius angaleti	Egg-larval Braconid parasite in Borneo
	Biosteres angaleti	Braconid larval-pupal parasite from the Indo-Australian Region
	Diachasmimorpha (= *Biosteres*) *dacusii*	Braconid parasite from India and Thailand
	= *Psytallia incisi* (= *Opius incisi*)	Braconid egg-larval parasite from SE Asia, est. in Hawaii, in lab and India
	Diachasma fullawayi	Braconid parasite from Hawaii
	Diashasma tryoni	Braconid parasite from Hawaii
	Dirhinus giffardii	Chalcid pupal parasite from Hawaii
	Fopius arisanus	from Malaysia and Thailand
Bactrocera cucurbitae (cont)	*Galeus silvestrii*	Diapriid parasite from Hawaii
	Ipobracon sp.	Braconid parasite from India
	Opius sp.	Braconid parasite from Ceylon
	Opius humilis	Braconid parasite from Hawaii
	Pachycrepoideus dubius = vindemiae	A Pteromalid parasite from Hawaii
	Parangrus osborni	Mymarid parasite from Hawaii
	Splangia cameroni	Pteromalid parasite from Hawaii
	Splangia hirta	Pteromalid parasite from Hawaii
	Tetrastichus giffardianus	Eulophid parasite from Hawaii
	Dirhinus anthracia = Dirhinus sp.	Chalcid parasite from S. India, imported to Ceylon, in Hawaii, Fiji. From Africa to Middle East, SE Asia, S. Pacific, Australia. Parasite of many fly families.
	Spalangia nigra	Pteromalid, pro larval-pupal parasite from Europe, N. America, pro imported into Hawaii. Many diverse hosts
	Spalangia cameroni	Pteromalid, pro larval-pupal parasite from Europe, Africa, Asia and S. America, imported to pacific many diverse hosts.

	Pachynuron vindemmiae	Pteromalid pupal parasite from Morocco
	Fopius (= Opius) vandenboschi	Braconid parasite from SE Asia
	Spalangia sp.	Pteromalid parasite from SE Asia
	Syntomosphyrum indicum	Eulophid parasite from SE Asia
	Spalangia philippinesis (= endius)	Pteromalid pupal parasite from Hawaii
	Opius compensans	Braconid parasite from India with 10-12% parasitism
	Dirhinus auratus	Chalcid parasite from Ceylon
	Dirhinusluzonensis	Chalcid parasite from India
	Spalangia afra	Pteromalid parasite
	Spalangia philippinesis	Pteromalid parasite
	Spalangia Stomoxysiae	Pteromalid parasite
	Philonthus sp.	Stanohylinid predator from South China
	Opius # 4	Braconid parasite from Borneo which prefers large 3rd instar larva
	Utetes bianchii	From Thailand
Batcrocera curvipennis	*Biosteres longicaudatus* var. *novocaledonicus*	An Indo-Australian Braconid larval parasite, now in Hawaii, Fiji, America, Caribbean, Florida from New Calefornia
	Opius fijiensis	Braconid egg-larval parasite from Australia, S. Pacific
Batcrocera decipiens		
Batcrocera decurtans		
Batcrocera depressa		
Batcrocera distinct		
Batcrocera diversa	*Psyttalia fletcheri*	A Braconid parasite from Thailand
	Psyttalia makii	A Braconid parasite from Thailand
	Utetes bianchii	A Braconid parasite from Thailand
Batcrocera dorsalis	*Diachasmimprpha (= Opius, = Biosteres) longicaudata*	An Indo-Australian Braconid 2-3rd stage larval parasite, now in Hawaii, Fiji, America, Caribbean, Florida parasitism 1.8%
	Diachasmimorpha albobalteata	A Braconid from Thailand

Source: Stibick (2004)

Pathogens : *Beauveria bassiana* is a facultative fungus to attack adult fruit flies.

❑❑❑

10
Chapter

Integrated Management of Fruit Flies

In order to minimize the indiscriminate and injudicious use of chemical pesticides, integrated pest management (IPM) has been enshrined as cardinal principle of plant protection in the overall crop protection programmes. As far as fruit flies are concerned, IPM is an eco-friendly approach for managing pest problems encompassing available methods and techniques of pest control such as cultural, mechanical, biological and chemical in a compatible and scientific manner. The greater emphasis has been given on biological control including use of bio-pesticides. IPM is also useful in reducing the pesticides residues in exportable agricultural commodities.

Mango like most of the fruit tree crops is usually attacked by two or three key pests, several secondary pests and a large number of occasional pests in localized areas where it is grown. Fruit flies are the serious pests of mango. There are 4000 species of fruit flies in India. Fruit flies especially *Bactrocera dorsalis* Handle cause considerable yield loss, which ranges from 31-86%. The importance of this pest has gone very high in view of the embargo imposed upon by USA and Japan on the import of mango fruits from India for the fear of introducing the Oriental fruit fly (*Bacrocera dorsalis*) along with mango shipment. It is therefore necessary to reopen this trade channel by evolving suitable control measures to meet the acceptable quarantine measures.

The management of fruit flies perspective mainly for *B. dorsalis* can be classified into two categories *viz.,* pre-harvest and post-harvest management practices. For small scale and local market crops pre harvest management is sufficient while for the export market a combination of pre and post harvest management practices are needed.

10.1. PREHARVEST MANAGEMENT PRACTICES

Fruit flies oviposit on fruits while in the field. Therefore, it is necessary to control this pest before fruits are harvested from the orchard. The following pre-harvest fruit fly control measures are followed for the effective management.

10.1.1. Cultural Control

Collection and proper disposal of the infested and dropped fruits prevent the further multiplication and carryover of population. Ploughing of the orchards during November-December exposes the diapausing pupae to sun's heat which kills them. Two to three times ploughing of fields reduces the pest population by exposing the pupa to the natural enemies also. Three weekly inter tree ploughing and raking under and below trees 45 days prior to harvest reduced the infestation of *B. dorsalis* in mango by 77-100% at Bangalore. Fruit fly population can be reduced by expanding planting distance in order to reduce humidity and increase solar radiation within orchards by eliminating fruit fly host plants in the hedges of mango orchards and by removing early or late fruits from the trees. Smoking in orchards is also reported to be helpful in preventing the fruit fly infestation. Early harvesting of fruits before ripening prevents egg laying by the adult fruit fly. Studies conducted at Pantnagar, UP, India during 1997 indicated that *B. dorsalis* damage was reduced to 4.6% by harvesting mango variety, Dashehari fruits at physiological maturity compared to 10% in fully ripe dropped fruits.

10.1.2. Crop Sanitation

Crop sanitation is simple and first measure of defence. Before harvest, ongoing efforts to pick up and destroy fallen fruits can reduce fruit fly infestation. It consists of regular destruction of fallen fruits, over ripe or damage in trees that are breeding grounds for fruit fly. The collected fruits may be either burnt or fed to pigs or placed in a tightly closed plastic bag under the sun for several days. Fruits may be buried, but must be covered by at least 50 cm of soil. Down remaining fruits from trees after harvest with a wooden or fiber glass pole.

10.1.3. Resistant varieties

Growing of tolerant/resistant varieties is the best way to save fruits from fruit fly damage as it is a completely hazardous proof method. The Dasheri and Rajapuri varieties of mango are reported tolerant to orchard flies in Gujarat, Screening of mango varieties at IIHR, Bangalore shows that Dasheri and Langra were least infested (3 and 4% respectively) by flies. The frequency of *Bactrocera* injury to physiologically mature fruit of Dasheri was less and highest damage was reported in fully ripe fruit of Mallika followed by Totapari.

10.1.4. Growing of Trap Crop

Tulsi, *Ocimum sanctam,* an aromatic plant containing 40% of Methyl eugenol attracts fruit fly males and acts as a sex attractant. Raising Tulsi plants in and around mango/fruit orchards and then spraying with insecticides (malathion, DDVP, fenthion etc.) help in fruit fly control.

10.1.5. Bagging

Fruit can be easily protected against fruit flies by bagging them in newspaper bags. The bag provides a physical protection to the fruit fly preventing adult female flies from laying eggs. The carambola export industry in Malaysia worth 10 million US $ protects entire orchards by bagging. It is also widely practicable to protect mangoes in Thailand and Philippines. Bagging is inexpensive and easy to apply and guarantees nearly complete protection from fruit flies. It is done at 55-60 days flower bloom or when the fruits are about the size of a chicken egg. Remove the bags when the fruit is mature. These bags are available in either waxed paper, cloth or news paper bags. In Southeast Asia newspaper is used to make the bag. The cloth bags are sturdy, washable, and can be used year after year. The fruit ripens fully within the bag. Pre-harvest wrapping of mango with brown paper resulted in at least 70% of fruit being of export quality compared with 50% of unbagged fruits.

10.1.6. Biological control

Despite large amounts of efforts devoted to the use of biological control agents to control fruit flies, there have been relatively few instances that may be regarded as sustainable successes. Generally, predators have little effect on fruit fly population in orchards. Invertebrate predators may include spiders, ants, carabid beetles, assassin bugs and other. Predation by fruit eating vertebrates such as birds and primates results in marked reduction in fruit fly numbers.

The use of parasitoids to control fruit flies has always had wide appeal. In general parasites are proved to be good targets for biological control. Inundative release of laboratory reared parasitoids of *Diachasmimorpha longicaudata, Fopius arisanus* and *F. vamdenboschi* resulted in 95% reduction of fruit flies in Hawaii. The solitary endoparasitic wasp, *Diachasmimorpha longicaudata* parasitizes larvae, when fly reaches to pupal stage, the wasp eggs hatch and larvae begin feeding on the fly pupa killing the fly before it can develop in to an adult. This is the classical biological control technique which showed 90% reduction in fruit fly population in Gainesville in 1990. Introduction of *Fapius arisanus* and *D. longicaudata* in to Pohnpei and Kosrae Islands in 1997 resulted in quick establishment of parasitoids.

10.1.7. Monitoring of fruit flies

For successful implementation of IPM programme, monitoring system usually involves some trapping device to sample the population. In case of fruit flies, they can be baited with either an olfactory attractant or may be trapped entirely on visual cues. A great deal of attention has been given by researchers to design trapping device for fruit flies. The trap design should be related to its intended use. The major purpose of a monitoring device is to take representative sample from population with accuracy and with the same efficiency during whole of the season. In this context, olfactory traps are more effective in terms of number of insects which they catch compared with yellow traps. (Economopoulous *et al.,* 1986). Fruit fly numbers are easily monitored with different parapheromone traps, *viz.*, Dakpot trap, protein hydrolysate trap, McPhail trap, Lynnfield trap, modified Steiner trap, Methyl eugenol trap (developed by IIHR, Bangalore) and Trimid lure trap. (Agrawal and Srivastava, 2008).

TABLE 9 : Particulars of trap type/lures against fruit fly species.

S. No.	*Target fruit fly species*	*Trap types*	*Lure/Attractant used*
1.	**Anastrepa* spp. And other not attracted by parapheromones	McPhail	Protein hydrolysate
2.	*Bactrocera* spp (mostly cucurbitae, *B. tryoni* and *B. tau*)	Lynfield/Steiner	Cue lure + malathion
3.	*Bactrocera* spp (mostly dorsalis complex)	Lynfield/Steiner	Methyl eugenol + malathion
4.	*Ceratitis capitate* and *C. rosae* males	Jackson/Steiner	Trimid/capilure + dichlorvos impregnated 1.5 cm sq discs
5.	*C. capitate* and *C. Rosae* females	Mc Phail	Protein hydrolysate

*Exotic fruit fly species of quarantine significance to India
Source : Drew (1974)

As the use of para pheromone traps leads to minimum environmental and ecological pollution, much emphasis has been given on the development of effective traps. In India, of the several pheromones reported, only methyl eugenol is readily available. Methyl eugenol is the most powerful among all the lures attracting *Bactrocera* spp. (Drew 1974, 1989, Drew and Hancock 1994). Earlier workers used extracts of *Ocimum sanctum* to study the population fluctuation in *B. correcta* (Reghunath and Indira 2000, Madhura, 2001).

10.1.8. Effect of colour of trap

Gowda (2018) selected three different experimental sites for evaluation of effect of different coloured traps on fruit flies. The maximum fruit flies were attracted to green coloured hand made bottle traps and minimum number of fruit flies were attracted to Rakshak trap used as standard. Most abundant species trapped in all types of coloured traps was *B. zonata* followed by *B. dorsalis, B. nigrotibialis* and the lowest species recorded was *B. correcta.*

10.1.9. Male annihilation technique (MAT)

The control of fruit flies by male annihilation technique (MAT) exploits the attraction of males to Parapheromone lures. It consists of a male lure combined with an insecticide (usally technical malathion and more recently fipronil), to reduce the male population of fruit flies to such a low level that mating does not occur. This is achieved by distributing cordelites (cotton strips), canite blocks, plywood blocks impregnated with the lure/ insecticide mixture. They are distributed from the ground or air in densities of at least 400/km^2. This treatment is repeated every 8 weeks.

Methyl eugenol has been successfully used in management programmes owing to its apparent olfactory action to which the species of *Bactrocera* are attracted. The methyl eugenol is known to attract flies from a distance of 800m. Methyl eugenol when used together with a pesticide and impregnated into a suitable substrate, forms the basis of MAT. The parapheromone + insecticide is used either as spray or placed in trap. The delivery systems used are cotton plug, alkathene foam, wooden blocks etc. Various traps used for attracting and killing the fruit flies are plastic jar trap, Mcphail trap, Patel fruit fly trap, Steiner trap, IMFFI trap and sticky trap etc. The bait consisting of Ethanol : Methyl eugenol : DDVP (6 : 4 : 1) incorporated in commercial plywood blocks (5 × 5 × 1-2 cm) placed in IMFFI trap is found most convenient, economical and effective. Installation of methyl eugenol incorporated blocks @ 10 blocks/ha on wide areas was very effective in North and South Gujarat.

10.1.10. Protein Bait Application Technique (BAT)

Protein bait spraying is an effective fruit fly control method. It consists of spraying a solution of protein and insecticides in water over a small area of leaves on each tree. Sexually immature and some mature female flies in need of protein meal to produce viable eggs are attracted to the bait, feed on it and are killed by the insecticide. One source of protein is Maurin Pinnacle Protein Insect Lure (MPPIL). Bait spraying is environmentally friendly because it specifically targets fruit flies, a minimum amount of insecticide is used, the fruits themselves are not sprayed and natural enemies are conserved as most of the foliage is not sprayed. The formulations consist of 50 ml of MPPIL concentrate + 4 ml of malathion 50% EC made up to one litre with water, sprayed at a rate of 50-100 ml on the under surface of one square metre of leaves on each tree. Spray are repeated every seven days starting one month before fruits starts maturing. The most commonly used protein now a day is a yeast autolysate consisting of very less salt and this reduces phytotoxicity. It is produced in Queens land under trade name Mauris Pinnacle Protein Insect Lure (MPPIL).

In Malaysia, it is marketed as Promar and Royal Tongalure in Tonga. To overcome the problem of bait being washed off from leaves and to improve the effectiveness of bait sprayer, Aventis Crop Science, (Australia) has developed a new formulation with fipronil instead of malathion and an additive that forms a gel that sticks to tree leaves and resists rain, New product known as Bactogel ™ has been used since late 1998 in the fruit fly eradication programme in Nauru and is available to countries that have registered fipronil insecticide.

10.1.11. Chemical control

10.1.11.1. Soil application : Application of chlorpyriphos, Lindane, and carbofuran etc. in soil followed by raking helps in destruction of pupae and reduced fruit fly infestation.

10.1.11.2. Foliar application : The insecticides posing no health hazard due to its toxic residue should be preferred for cover spray particularly during fruiting. Cover sprays using malathion, quinalphos, carbaryl, imidacloprid, DDVP, neem oil etc at recommended dose are effective. Two insecticidal sprays *viz* carbary 1 0.2% at 30 days prior to harvest and decamethrin 0.0028% at 15 days prior to harvest gave good control of pest. Application of bait sprays within 3-4 days of trap catches *viz* deltamathrin 0.0025% + 0.1% molasses and after 10-12 days, fenthion 0.05% + 0.1% molasses followed by dimethoate 0.045% + 0.1% molasses are found effective.

10.1.12. Sterile Insect Technique (SIT)

One of the most effective and environmentally sound methods for eradication of insect pest is the sterile insect technique (SIT) pioneered by Knipling (1955). SIT has been demonstrated to be an effective tool for suppression and eradication of a number of species of tephritid flies, particularly when coupled with other control techniques. Control is achieved in SIT by mass release of sterile males which mate with wild females. Females that mate with the sterile males do not produce viable eggs and over a time this results in population decline and possibly eradication. Although it is an effective population management tool, it is expensive in terms of both money and time.

The following pre harvest standardized IPM, combining sanitation, interculture ploughing + three insecticidal sprays at 15 days interval prior to harvest, will give 95-99% control :

10.1.13. Pre-harvest IPM (Starting 45 days prior to harvest)

(i) Destroy all fallen fruits at weekly interval

(ii) Rake/plough between trees once or twice.

(iii) Sprays the fruits at :

— 45 days prior to harvest by carbaryl (0.2%)

— 30 days prior to harvest by Dimethoate (0.06%)

— 15 days prior to harvest by Dimethoate (0.0028%)

(iv) Avoid delay in harvest.

10.2. POST HARVEST CONTROL MEASURES

Fruits that are harvested may carry eggs and grubs, leading to rotting. Therefore, post harvest treatments are very much necessary to prevent rotting and thereby economic loss.

10.2.1. Heat treatment techniques

Hot water immersion, high temperature forced air and vapour heat are three heat treatment techniques that can be used for post harvest insect control for perishable commodities like fresh fruits, fresh vegetables, bulbs and cut flowers. Heat treatment waned with the development of chemical fumigants, which could be applied cheaply and easily. Today with increasing cost of development of developing new chemical and regulatory restrictions on existing ones, interest in heat disinfection has been revived.

Heat treatment technologies are currently relatively simple non-chemical alternative that can kill quarantine pests in perishable commodities as well as control some post-harvest diseases. Heat treatments do not pose significant health risks and are more appealing to consumers. In most cases heat treatments are performed by the country of origin before a product is exported. The temperature, duration and application method is both cultivar and commodity specific and must be very precise to kill pests without damaging the commodity. Fruit response to heat varies depending on the condition of the fruit fly to treatment, the commodity concerned, the temperature and duration of treatment as well as mode of heat application (*i.e.* hot air vs water). If not properly applied heat treatments may result in commodity damage which typically is manifested as browning fruit surface, uneven ripening and breakdown of the fruit flesh. Heat treatment does not affect market quality of the commodity.

(a) Hot water treatment : It consists of submerging the commodity in hot water bath at a specific temperature for specified time based on the commodity being treated and the pests that may be present in perishable food commodities. The mandated probit 9 level of fly control can be achieved by treating the core of the fruit to 43°C-46.7°C with exposure time varying from 35-90 min. Variations are noted for different commodities pest species, and life stages of insect pests. Hot water is an effective heat transfer medium and when properly circulated through the load of fruit, quickly establishes a uniform temperature profile. Hot water immersion is currently used to successfully treat mangoes infested with Mediterranean fruit fly and several different species of *Anastrepha* before importation in to US from Mexico.

Double dip method or two stage hot water immersion treatment involves the immersion of mango fruit in water at 40°C for 20 minutes, followed by 10 minutes at 46°C to get 100 per cent mortality of *B. dorsalis* eggs. The quarantine treatment with hot water some time (46°C for 90 minutes) affects fruit epicuticular waxes, fruit ripening and weight. Hence use of edible waxes (fatty acids 2% + carbohydrates 0.5% + calcium caseinate 0.1%) after hot water treatment showed positive response on maturity. Hot water treatment (48-50°C for 20 min) followed by fumigation with EDB (12 g/m^3 for 2 hrs at 20°C) and storage at 7-10°C for 4 days resulted in no survivors of *D. dorsalis.*

(b) High temperature forced air treatment : Re-circulated air that has been heated and humidified can be forced over fruit surfaces to raise the temperature to a level that is lethal to target pest species. Heated air treatment of 40-50°C (usually at 4 incrementally increased temperatures) for less than 8 hours is becoming more common for fruit fly control in tropical commodiries. Condensation on fruit surfaces or in the treatment chamber is prevented by keeping the dew point temperature 2-3°C below the dry bulb temperature throughout the duration of the test. This precise control of temperature and RH is advantageous because it prevents fruit desiccation and scalding.

(c) Vapour-Heat treatment (VHT) : It uses heated air saturated with water vapour to heat perishable food commodities to a specified temperature and hold the temperature for a specified period to ensure that all pests (such as tephritid fruit immatures) within the commodity are killed. Typically the pulp temperature of the commodity is the saturated water vapour to 43.3-44.4°C during a period of 6 or 8 hr. Vapour heat (greater than 90% RH) is approved by USDA-APHIS for treatment of mangoes.

(d) Fumigation : Fumigation of fruit was most commonly used for controlling quarantine pests on perishable commodities at specific temperature and doses. Methyl bromide and ethylene di-bromide were formerly used by many countries for controlling the pests. These chemicals have been banned by many countries for fear of toxic residues on the treated fruits. Hence this method is not in practice now a days.

Hot water immersion system can be easily assembled and is durable, mobile and inexpensive and more efficient than vapour heat. However, it can damage to some extent fruits and vegetables during treatment. It is the only approved quarantine treatment for mangoes. Alternatively, vapour heat and forced air treatment systems are less damaging to the commodities and more versatile then other treatment systems, however, they are more expensive.

10.2.2. Irradiation as quarantine treatment

In 1986 the Food and Drug Administration (FDA) approved the use of radiation treatment of up to KGY (100 Krad) on fruits and vegetables. Irradiation is effective in killing, sterilizing or preventing further development of a wide variety of insect pests of quarantine importance on perishable fruits and vegetables. With the threat of an impending loss of methyl bromide, a fumigant very commonly used for quarantine control of insect pests, there has been a resurgence of interest in irradiation as a quarantine treatment. Radiation may be provided by gamma rays from cobalt 60 or cesium 137 sources, electors are generated from machine sources or by x-rays. For fruit fly pests quarantine security for a single treatment protocol will be defined as the prevention of adult emergence at probit 9 (99.9968% mortality). A dose of 150 GY is recommended as a generic quarantine treatment against potential infestation of mango fruit flies in exported mangoes. It was found that a dose of 0.5 KGY killed most resistant stage of fruit fly *D. dorsalis* in Philippines. Gamma irradiation at 75 GY is the only successful treatment acceptable by department to accept consignments with no viable insects in Australia.

Following Post Harvest Management is recommended :

(i) Post harvest hot water treatment, within 24 hr after harvest at 48°C is maintained by thermostate for one hour.

(ii) 100% fruit fly infestation free fruits were obtained by hot water treatment in 5% salt solution at 55°C for 30 min. without thermostat gave 100% control (Annual Report IIHR, 1998-99).

CONCLUSION

Keeping in view the importance of the pest and crop, most of the fruits can be managed in the grower's field using any of the option combinations available including field sanitation, bagging of fruits, use of methyl eugenol traps, spray of protein bait with toxicants, growing fruit fly resistant genotypes, augmentative release of biological control agents and use of soft insecticides. On the other hand, incorporation of a number of different techniques including the sterile insect technique, quarantine and post harvest heat treatment could be exploited for better results in wide area management of fruit fly.

❑❑❑

11
Chapter

Mass Trapping Technique

Fruit flies nowadays are the main constraints to production and marketing of mango and guava fruit in India. The pre and post harvest yield loss due to fruit fly of mango and guava ranges from 27-80% (as fruit drop and fruit pulp damage) and 20-46% respectively (Haseeb, 2007). Trapping fruit flies is one of the alternatives to mango pest in an ecofriendly way. These traps have high specificity, low cost and are environmentally quite safe (Sureshbabu and Viraktamath, 2003). Among alternative methods of fruit flies control in fleshy fruit, male annihilation technique (MAT) and bait application technique (BAT), seem to have great potential. Food attractants provide an effective system for capturing the fruit fly as BAT. Methyl Eugenol (ME) a strong male attractant of fruit flies, is being used as MAT. The system would lead to more effective population suppression and provide a better monitoring tool by attracting males and females in area-wide management of fruit flies. Therefore, there is worldwide requirement for acceptable suppression measures capable of maintaining population at very low levels and ultimately employed for fruit fly eradication programmes. Some important suggestions and recommendations are given as under for mass trapping of fruit flies :

- The trap should be placed about 1.5 m above ground and should not be exposed to direct sunlight wind and dust.
- Trap entrances should be cleared of leaves to allow access to flies and prevent entry of ants.

- The wire hanging the trap should be coated with a sticky ant barrier or Vaseline for further protection against ants.
- Ten traps may be hanged in a hectare for mass trapping of fruit flies.

Thees traps can be prepared by empty plastic bottles of cold drink or mineral water. In each traps charging with alcohol, methyl eugenol and malathion (50 EC) prepared in the ratio of 6 : 4 : 1 is done. The traps can be fabricated of 1.5 lit size and 1lit capacity with three windows. Wooden piece of 5 × 5 × 0.5 cm are charged with 6 : 4 : 1 (alcohol : methyl eugenol : malathion) by soaking for one week and then shade dried. These are placed in the loop of plastic wire and replaced every month (Fig. 67, 68, 69).

Similar Rakshak trap of 1lit capacity is transparent with three windows of equal size. Cotton wicks with string on the lid are also charged with 4 : 0.05 Methyl eugenol : malathion. These cotton wicks are recharged every month.

11.1. PREPARATION OF TRAPS

The traps used can be fabricated in cylindrical and square shape or Rakshak traps recommended by Agricultural and Processed Food Products Export Development Authority (APEDA), Ministry of Commerce and Industry, Government of India and can be obtained from the market. The cylindrical traps can be made from different soft drink bottles of 1.5 lit capacity. These bottles are cut at the neck and that cut piece was reversed on the top and the plastic wire is passed from the top to the bottom to hold the 5 × 5 × 0.5 cm wooden piece. Three windows are cut in the bottle for entry of the fruit flies.

Rakshak trap is one litre plastic container (Fig. 66) with 10 cm base, 9 cm diameter top and 11.5 cm deep. It has a yellow coloured screw top lid. Four entry holes of 2.5 cm in diameter evenly spaced 1.5 cm below the lid are located in trap. A plastic chain is hooked in the centre on the upper side of screw top. On the lower side of the lid there is a hook for holding the wick/plywood.

TABLE 10 : Concentration of lure and insecticide to each trap :

Lure mixture	*Volume*	*Concentration*
*Cue lure + malathion (CUE)	4 ml	2 ml applied to each side of a 4 wick clip
*Methyl eugenol + malathion (ME)	4 ml	2 ml applied to each side of a 4 wick clip
Trimid/Capi lure	4 ml	2 ml applied to each side of a 4 wick clip

*Cue lure mixture : 125 ml UL malation to 1 litre cue lure

*Methyl eugenol lure mixture : 125 ml UL malathion to 1 litre methyl eugenol

11.2. PREPARATION OF COTTON WICK AND WOODEN PIECE

Absorbent cotton wool is used for wick preparation. After taking-out required quantity of cotton from the pack, the same is put in between two palms and rolled with a light pressure to make a wick. From one pack weighing 250 g about 833-834 wicks could be

FIG. 66. Rakshak Trap

FIG. 67. Fruit fly trap made from cold drink bottle

prepared *i.e.* about 0.3 g cotton is required to prepare one wick. The wooden piece of 5 × 5 × 0.5 cm is cut with the help of cutter and the hole is made in the middle of the piece to hold the thread.

FIG. 68. Fabricated trap

FIG. 69. Fabricated trap

11.3. PREPARATION OF ATTRACTANT MIXTURE FOR WOODEN PIECES

Solution is prepared by taking alcohol, methyl eugenol and malathion in the ratio of 6 : 4 : 1. To make the solution, 600 ml of alcohol, 400 ml of methyl eugenol and 100 ml of malathion (50 EC) is mixed. The wooden pieces are dipped in this mixture for one week.

11.4. CHARGING OF COTTON WICK

Fresh mixture of Methyl eugenol + Malathion 50 EC is prepared in ratio of 4 : 0.05 and kept in screw capped glass container. Then 4 ml of mixture is taken with the help of disposable syringe of 5 ml capacity and injected in the wick already hanged in the trap. Mixture of Methyl eugenol (100 ml) + Malathion 50 EC (1.25 ml) is stored in screw capped glass container. Then 4 ml of mixture is taken with the help of disposable syringe of 5 ml capacity and injected in the wick already hanged in the trap.

The wooden pieces are taken out from the hanging traps after one month and recharged by dipping in mixture. The charging of wick and wooden piece is done after one month.

❑❑❑

Chapter 12

Taxonomical Characters of Major Fruit Flies in Northern India

12.1. CHARACTERS OF ORIENTAL FRUIT FLY *BACTROCERA DORSALIS* (HENDEL)

Body

Body predominantly black or dark fuscous or a balanced mixture of black and yellow.

Head

(*i*) Number of frontal bristles two pairs

(*ii*) Number of optical bristles one pair.

(*iii*) Anterior orbital bristle of male normal, unmodified.

(*iv*) Posterior orbital bristles reclinate, accumulate.

(*v*) Ocellar bristles absent or minute.

(*vi*) Post ocullar bristles present but minute.

(*vii*) Outer vertical, post-vertical and postocular bristles all acuminate.

(*viii*) Postocular bristles acuminate, reduced in number and size both.

(*ix*) Genal bristles present.

(*x*) Face undulating, convex above and concave below with transverse sulcus, or with distinct antennal grooves and carina, about as long as frons vertical.

(*xi*) Dark spots in antennal furrows.

(*xii*) Antennal spots are transverse,

(*xiii*) Parafacial spots absent.

(*xiv*) Frons and parafacial without small silvery markings.

(*xv*) Antenna about as long as face or considerable longer than face.

(*xvi*) Arista longer than first flagellomere, bare or with hairs distinctly shorter than greatest aristal width.

(*xvii*) Probosis short and capitate.

Thorax

(*i*) Inner scapular bristle present and distinguishable from surrounding vestiture, dark.

(*ii*) Outer scapular bristle present and distinguishable from surrounding vestiture, dark.

(*iii*) Post pronotal bristles absent.

(*iv*) Presutural dorsocentral bristle absent.

(*v*) Presutural supra-alar bristle absent.

(*vi*) Postsutural supra-alar bristles present, well developed, similar to post alar bristle.

(*vii*) Intrapost alar bristle absent.

(*viii*) No. of scutellar bristles one pair.

(*ix*) The single pair of scutellar bristles apical.

(*x*) Anterior notopleural bristle present.

(*xi*) Anepisternal bristle dark, brown to black.

(*xii*) Long, erect setulae on leterotergite absent.

(*xiii*) Scutal setulae acuminate and pale.

(*xiv*) Scutellum densely setalose.

(*xv*) Setulae on scutellum short, decumbent, unicolours acuminate.

(*xvi*) Transverse suture with the lateral branches wide apart.

(*xvii*) Scutum black with orange to red brown areas, or black without a large dark central strip broadens basally.

(*xviii*) Post pronotal lobe entirely pale whitish.

(*xix*) Dark lyre like pattern on scutum absent.

(*xx*) Discrete shiny black spots on scutum absent.

(*xxi*) Median longitudinal black strip on scutum absent.

(*xxii*) Number of pale whitish to yellow postutural stripes two (lateral), also called lateral vittae.

(*xxiii*) Area bordering scutoscutellar surface medially with dark brown, light coloured or without dark brown spot, light coloured or without dark brown spot.

(*xxiv*) Distinct pale vertical strip extending half way between posterior half way between posterior half of notopleuron and anterior notopleural bristle.

(*xxv*) Scutum microtrichia in discrete pattern due to density differences.

(*xxvi*) Dorsum of scutellum flat or slightly convex not swollen.

Legs

(*i*) Femora slender with yellowish colour.

(*ii*) Forefemur with regular bristles; without ventral spines; with 1 to 3 posterodorsal and one posteroventral rows of bristle only.

(*iii*) Mid femur and hind femur without spine like bristles.

(*iv*) Middle leg of male without feathering.

(*v*) Femora all entirely yellow colour, dark mark on fore femur.

Wings

(*i*) Wing partly bare.

(*ii*) Cell bc microtrichia absent.

(*iii*) Cell c microtrichia present in apical area.

(*iv*) Cell dm entirely microtrichose.

(*v*) Dense microtrichia at end of vein Al + Cu A2 in male present.

(*vi*) Dominant wing pattern costal banded.

(*vii*) Wing pattern mostly brownish.

(*viii*) Dark longitudinal streaks through basal cells absent.

(*ix*) Costal band coloured from Sc to beyond R_{4+5}.

(*x*) Apex of costal band not distinctly expended.

(*xi*) Costal band not extending below R_{2+3}, or extending below (very slightly).

(*xii*) All one coloured.

(*xiii*) Crossvein r-m hyaline.

(*xiv*) Crossvein dm-cu hyaline.

(*xv*) Anal band present, reaching nearly to wing margin.

(*xvi*) Ratio of width of apical band in cell R_{4+5} to length of r-m 0.25-0.33.

(*xvii*) Costal spines at subcostal break absent.

(*xviii*) Distance between crossvein r-m and costal equal to r-m to shorter than r-m.

Abdomen

(*i*) Abdomen ovate.

(*ii*) Abdominal tergitis separate.

(*iii*) Abdomen in lateral view, arched, domelike, rigid; abdominal tergite broader at apex than at base.

(*iv*) Pecten of dark bristles on tergite 3 of male present.

(*v*) Tergal gland on tergite 5 present.

(*vi*) Abdominal tergite 5 normal.

(*vii*) 6th tergite of female normally concealedy shorter than 5th.

(*viii*) Abdominal setulae acuminate and pale.

(*ix*) Abdominal micromentum absent.

(*x*) Posterior margin of sternite 5 of male with deep V-shaped posterior concavity.

(*xi*) Abdominal tergites 3-5 predominantly yellow to orange brown.

(*xii*) Abdominal tergites with medial dark strip on T3-T5 and a transverse dark like on T3; not brown with medial T shraped yellow mark.

(*xiii*) T-shaped transverse band present on last 3 Abdominal segments.

12.2. CHARACTERS OF ADULT PEACH FLY, *B. ZONATA* (SAUNDERS)

Body

Body predominantly orangish to brown.

Head

(*i*) Number of frontal bristle two pairs.

(*ii*) Number of orbital bristle one pair.

(*iii*) Anterior orbital of male normal, unmodified.

(*iv*) Posterior orbital bristle reclinate.

(*v*) Ocellar bristles absent or minute, like setulae.

(*vi*) Postocellar bristles absent.

(*vii*) Postocular bristles acuminate (normal).

(*viii*) Genal bristle present.

(*ix*) Head higher than long.

(*x*) Face undulating, convex above, concave below with distinct atennal grooves, dark spots in antennal furrows (small round).

(*xi*) Parafacial spot absent.

(*xii*) Frons and parafacial without silvery markings.

(*xiii*) Eye round, about as high or slightly higher than long.

(*xiv*) Antenna considerably longer than face.

(*xv*) Arista longer than flagellomere, bare or with hairs.

(*xvi*) Proboscis short, capitate.

Thorax

(*i*) Inner scapular bristle absent.

(*ii*) Outer scapular bristle present.

(*iii*) Post pronotal bristles present.

(*iv*) Presutural dorsocentral bristle absent.

(*v*) Number of scutellar bristles one pair.

(*vi*) The single pair of scutellar bristles apical.

(*vii*) Anterior notopleural bristle present.

(*viii*) Posterior notopleural bristle acuminate.

(*ix*) Number of outstanding anepisternal bristles one.

(*x*) Katepisternal bristles absent.

(*xi*) Anepisternal bristles pale, yellowish.

(*xii*) Long, erect setulae on laterotergit absent.

(*xiii*) Scutum orange-brown, or red brown, without a large dark central strip which broadens basally.

(*xiv*) Post pronotal lobe entirely pale whitish, concolourus with lateral postsutural stripe (vittae).

(*xv*) Dark lyre like pattern on scutum absent.

(*xvi*) Discrete shiny black spots on scutum absent.

(*xvii*) Median longitudinal black stripe on scutum absent.

(*xviii*) Number of pale whitish, postsutural strips (lateral vittae) two.

(*xix*) Transverse suture without distinct stripe or spot.

(*xx*) Dorsum of scutellum flat or slightly convex, not swollen.

(*xxi*) Scutellum normal, without a dark and pale pattern.

Legs

(*i*) Femora slender with yellowish white colour.

(*ii*) Fore femur with regular bristles, without ventral spines.

(*iii*) Middle femur and hind femur without spine-like bristles.

(*iv*) Middle leg of male without feathering.

(*v*) Femora all entirely of one colour.

Wings

(*i*) Wing partly bare.

(*ii*) Cell dm bare on basal one-third to one-half.

(*iii*) Dense microtrichia at end of vein Al + Cu + Cu A2 in male present.

(*iv*) Dominant wing pattern costal-banded.

(*v*) Wing pattern mostly yellowish brown.

(*vi*) Costal band with only cell Sc and apex of vein R_{4+5} coloured.

(*vii*) Apex of costal band distinctly expended into a spot.

(*viii*) Costal band not extending below R_{2+3}, pale.

(*ix*) Cross vein r-m hyaline.

(*x*) Cell r_{2+3} apical to r-m with large hyaline area.

(*xi*) Cell r 1 and r_{2+3} with distinctly darker spots within dark area pattern.

(*xii*) Intercalary band absent.

(*xiii*) Sub-costal crossband absent.

(*xiv*) Posterior apical crossband absent.

(*xv*) Outstanding costal spines at subcostal break absent.

(*xvi*) Distance between crossvein r-m and costa shorter than r-m.

(*xvii*) Cell dm widens apically gradually from base.

Abdomen

(*i*) Abdomen ovate.

(*ii*) Abdominal tergites separate.

(*iii*) Abdomen in lateral view, arched, domelike, rather tergit.

(*iv*) Pecten of dark bristles on tergite 3 of male present.

(*v*) Tergal gland on tergit 5 present.

(*vi*) Abdominal tergite 5 normal.

(*vii*) 6th tergite of female normally concealed, shorter than 5th.

(*viii*) Abdominal setulae acumulate and pale.

(*ix*) Posterior margin of sternite 5 of male with deep V-shaped posterior concavity.

(*x*) Abdominal tergites 3-5 predominantly yellow to orange brown.

(*xi*) Abdominal tergites with medial dark stripe on T3-T5 and a transverse dark line on T3 with medial dark stripe on T5 or without medial dark stripes, not brown medial T shaped yellow mark on abdomen.

12.3. CHARACTERS OF ADULT GUAVA FRUIT FLY, *B. CORRECTA* (BEZZI)

Body

Body a balanced mixture of black and yellow.

Head

(*i*) Number of frontal bristle two pairs.
(*ii*) Number of orbital bristle one pair.
(*iii*) Anterior orbital of male normal, unmodified.
(*iv*) Posterior orbital bristle reclinate, acuminate.
(*v*) Ocellar bristles absent or minute, like setulae.
(*vi*) Postocellar bristles present (minute).
(*vii*) Postocular bristles present (minute).
(*viii*) Outer vertical, postvertical and postocellar all acuminate.
(*ix*) Post ocular bristles acuminate, normal.
(*x*) Genal bristle present.
(*xi*) Head higher than long.
(*xii*) Face undulating, convex above, concave below with transverse sulcus, shorter than frons vertical.
(*xiii*) Parafacial spot absent.
(*xiv*) Frons and parafacial without silvery markings.
(*xv*) Antenna considerable longer than face.
(*xvi*) Arista short, capitate.
(*xvii*) Antennal spots with round shape.

Thorax

(*i*) Inner scapular bristle present and distinguishable from surrounding vestiture, dark brown.
(*ii*) Outer scapular bristle present and distinguishable from surrounding vestiture, dark brown.
(*iii*) Post pronotal bristle absent.
(*iv*) Presutural dorsocentral bristle absent.
(*v*) Presutural supra-alar bristle absent.
(*vi*) Post sutural supra-alar bristle present, and well developed.
(*vii*) Number of scutellar bristles one pair.
(*viii*) The single pair of scutellar bristles apical.
(*ix*) Anterior notopleural bristle present.
(*x*) Posterior notopleural bristle acuminate.
(*xi*) Scutellum densily setulose.
(*xii*) Scutum black with orange to read brown areas or black.
(*xiii*) Post pronotal lobe entirely pale whitish.

(*xiv*) Posterior half of notopleuron pale whitish to yellowish.

(*xv*) Dark lyre like pattern on scutum absent.

(*xvi*) Discrete shiny black spots on scutum absent.

(*xvii*) Number of pale whitish postsutural strips (vittae) two (lateral).

(*xviii*) Subscutellum uniformly yellowish to pale orange brown or unformally brown or black.

(*xix*) Dorsum of scutellum flat or slightly convex, not swollen.

Legs

(*i*) Femora slender.

(*ii*) Fore femur with regular bristles, without ventral spines.

(*iii*) Mid femur and hind femur without spine-like bristles.

(*iv*) Middle leg of male without feathering.

(*v*) Femora all entirely of one colour, dark mark on fore, middle and hind femur.

Wings

(*i*) Wing partly bare.

(*ii*) Cell bc microtrichia absent.

(*iii*) Costal band discontinuous, breaking at cell R_{2+3}.

(*iv*) Cell dm bare on basal one third to one half or bare on less than basal one third.

(*v*) Dense microtrichia at end of vein Al + Cu A2 in male present.

(*vi*) Dominant wing pattern costal-banded.

(*vii*) Wing pattern mostly brownish.

(*viii*) Costal band with only cell Sc and apex of vein R_{4+5} coloured.

(*ix*) Cross vein r-m hyaline.

(*x*) Cross vein dm-cu hyaline.

(*xi*) Anal band present, reaching nearly to wing margin.

(*xii*) Subbasal crossband absent.

(*xiii*) Distance between crossvein r-m and costa shorter than r-m.

(*xiv*) Vein M distally straight.

Abdomen

(*i*) Abdomen ovate or parallel sided.

(*ii*) Abdominal tergitis separate.

(*iii*) Abdomen in lateral view, arched, domelike, rather rigid.

(*iv*) Pecten of dark bristles on tergite 3 of male present.

(*v*) Andominal tergite 1 broader at apex than at base, without a prominent hump laterally.

(*vi*) Tergal gland on tergits 5 present.

(*vii*) Abdominal tergite 5 normal.

(*viii*) 6th tergite of female normally concealed, shorter than 5th.

(*ix*) Abdominal setulae acuminate and pale.

(*x*) Posterior margin of sternite 5 of male with deep V-shaped posterior concavity.

(*xi*) Abdominal tergites 3-5 predominantly yellow to brown.

(*xii*) Abdominal tergites with medial dark strip on T3-T5 and a transverse dark line on T3, not brown with medial T-shaped yellow mark.

12.4. CHARACTERS OF ADULT *B. DIVERSA* (COQUILLETT)

Body

Body a balanced mixture of black and yellow.

Head

(*i*) Number of frontal bristle one pair.

(*ii*) Number of orbital bristle one pair.

(*iii*) Anterior orbital of male normal, unmodified.

(*iv*) Posterior orbital bristle reclinate, acuminate.

(*v*) Ocellar bristles absent or minute, like setulae.

(*vi*) Postocular bristles acuminate reduced in number and size, hardy detectable 25 × (pale).

(*vii*) Genal bristle present.

(*viii*) Head higher than long.

(*ix*) Face concave or undulating, convex above, concave below without transverse sulcus, antennal grooves, or carnia shorter than frons.

(*x*) Para facial spots absent.

(*xi*) Frons and parafacial without small silvery markings.

(*xii*) Eye round, about as high or slightly higher than long.

(*xiii*) Antenna considerable longer than face.

(*xiv*) Probosis short and capitate.

Thorax

(*i*) Inner scapular bristle present and distinguishable from surrounding vestiture, dark.

(*ii*) Post notal bristle absent.

(*iii*) Presutural dorsocentral bristle absent.

(*iv*) Presutural supra-alar bristle absent.

(*v*) Post sutural supra-alar bristle present.

(*vi*) Post sutural dorsocentral bristle absent

(*vii*) Anterior bristle present, well developed, similar to post-alar bristle.

(*viii*) Intra post alar bristle absent.

(*ix*) Number of scutellar bristles one pair or two pairs.

(*x*) Anterior notopleural bristle present.

(*xi*) Posterior notopleural bristle acuminate.

(*xii*) Anterior notopleural bristle present.

(*xiii*) Scutal setulae acuminate and pale.

(*xiv*) Scutellum densly setulose.

(*xv*) Scutum black; without a large dark central strip which broadens usually.

(*xvi*) Post pronotal lobe entirely yellowish.

(*xvii*) Dark lyre like pattern on scutum absent.

(*xviii*) Discrete shiny black spots on scutum absent.

(*xix*) Median longitudinal black strip on scutum absent.

(*xx*) Number of pale whitish postsutural strips two (lateral) (medial stripe sometimes absent in female) or three.

(*xxi*) Pale medial post sutural strip of scutum linear.

(*xxii*) Distinct pale vertical anepisternal stripe expending halfway between posterior half of notopeuron and anterior notopleural bristle.

(*xxiii*) Mediotergite uniformly brown or uniformly black or yellowish to orange brown medially with distinct dark spots laterally.

(*xxiv*) Dorsum of scutellum flat or slightly convex not swollen.

(*xxv*) Scutellum normal; without a dark and pale pattern.

Legs

(*i*) Femora slender.

(*ii*) Fore femur with regular bristles, without ventral spines.

(*iii*) Mid femur and hind femur without spine-like bristles.

(*iv*) Middle leg of male without feathering.

(*v*) Femora at least one femur markedly darken in apical part than in basal part, dark mark on fore femur, middle femur and hind femur.

Wings

(*i*) Wing partly bare.
(*ii*) Cell bc microtrichia absent.
(*iii*) Dense microtrichia at end of vein Al : cu + cu A2 in male absent.
(*iv*) Dominant wing pattern costal-banded.
(*v*) Wing pattern mostly brownish.
(*vi*) Cross vein r-m hyaline.
(*vii*) Cross vein dm-cu infuscated at posterior end (indistinct).
(*viii*) Anal band present, reaching nearly to wing margin, alone cell cup extension.
(*ix*) Subbasal crossband absent.
(*x*) Anterior apical band or costal band not extended to vein M.
(*xi*) Posterior apical cross band absent.
(*xii*) Outstanding costal spines at subcostal break absent.
(*xiii*) Vein m distally straight.

Abdomen

(*i*) Abdomen ovate or parallel sided.
(*ii*) Abdominal tergitis separate.
(*iii*) Abdomen in lateral view, arched, domelike, rather rigid.
(*iv*) Pecten of dark bristles on tergite 3 of male present.
(*v*) Tergal gland on tergits 5 present.
(*vi*) Abdominal tergite 5 normal.
(*vii*) 6th tergite of female normally concealed shorter than 5th.
(*viii*) Abdominal setulae acuminate and pale.
(*ix*) Abdominal tergites with medial dark strip on T3-T5 and a transverse dark line on T3 (also has transverse bands), or with medial dark stripe, usually in T3-T5 m or without medial dark strip, not brown with medial T-shaped yellow mark.

12.5 CHARACTERS OF ADULT MELON FLY, *B. CUCURBITAE* (CONQUILLETT)

Body

Body predominantly orange to browm

Head

(*i*) Number of frontal bristle three or four pairs.

(*ii*) Number of orbital bristle one pair.

(*iii*) Anterior orbital of male normal, unmodified.

(*iv*) Posterior orbital bristle reclinate, acuminate.

(*v*) Postocular bristles all acuminate (pale), normal.

(*vi*) Genal bristle present.

(*vii*) Head higher than long.

(*viii*) Face undulating, convex above, concave below with transverse sulcus.

(*ix*) Dark spots in antennal furrows.

(*x*) Parafacial spot absent.

(*xi*) Frons and parafacial without small silvery markings.

(*xii*) Eye round and slightly higher than long.

(*xiii*) Antenna longer than face.

(*xiv*) Probosis short and capitate.

Thorax

(*i*) Inner scapular bristle present and distinguishable from surrounding vestiture and dark.

(*ii*) Postpronotal bristle absent.

(*iii*) Presutural dorsocentral bristle absent.

(*iv*) Presutural supra-alar bristle absent.

(*v*) Post sutural supra-alar bristle present.

(*vi*) Anterior bristle present, well developed, similar to post-alar bristle.

(*vii*) Intra post alar bristle absent.

(*viii*) Number of scutellar bristles one pair.

(*ix*) Posterior notopleural bristle acuminate.

(*x*) Anepisternal bristle one, while kalepisternal bristle absent.

(*xi*) Posterior half of notoplearon yellowish white.

(*xii*) Dark lyre like pattern on scutum absent.

(*xiii*) Number of pale whitish to yellow postsutural strips Three.

(*xiv*) Pale medial post sutural strip (median vittae) on scutum linear.

(*xv*) Lateral post sutural strips of secutum extending to intra alar bristle or beyond.

(*xvi*) Scutum without balkish dorsocentral stripe.

(*xvii*) Discrete pale horizontal stripe along upper anepistemum absent or indistinct.

(*xviii*) Ketatergit with pale yellowish spot present and distinct.

(*xix*) Anatergite with pale yellowish spot present and distinct.

(*xx*) Subscutellum uniformly brown.

(*xxi*) Mediotergit uniformly yellowish to orange brown or uniformly brown.

(*xxii*) Dorsum of scutellum flat or slightly convex not swollen.

Legs

(*i*) Femora slender.

(*ii*) Fore femur with regular bristles, without ventral spines, with 1-3 rows of bristle only.

(*iii*) Mid femur and hind femur without spine-like bristles.

(*iv*) Middle leg of male without feathering.

(*v*) Femora all entirely of one colour.

Wings

(*i*) Wing partly bare.

(*ii*) Cell bc, microtrichia absent.

(*iii*) Cell dm bare on less than basal one third.

(*iv*) Dense microtrichia at end of vein Al + Cu A2 in male absent.

(*v*) Dominant wing pattern cross-banded, or costal banded with large apical spot.

(*vi*) Wing pattern mostly brownish.

(*vii*) Dark longitudinal streaks through basal cell absent.

(*viii*) Costal band coloured extended from Sc to beyond R_{4+5}.

(*ix*) Apex of costal band distinctly expanded into spot.

(*x*) Cross vein r-m covered by a major crossband infuscation.

(*xi*) Anal band present, reaching nearly to wing margin, along cell cup extension.

(*xii*) Subbasal crossband absent.

(*xiii*) Vein R_{2+3} generally straight.

(*xiv*) Distance between crossvein r-m and costa shorter then r-m.

(*xv*) A predominant brown infuscation on the crossvein dm-cu and a slight infuscation along crossvein r-m.

Abdomen

(*i*) Abdomen ovate or parallel sided.

(*ii*) Abdominal tergites separate.

(*iii*) Abdomen in lateral view, arched, domelike, rather rigid.

(*iv*) Abdominal tergit 1 broader at apex than the base.

(*v*) Pecten of dark bristles on tergite 3 of male present.

(*vi*) Tergal gland on tergits 5 present.

(*vii*) Abdominal tergite 5 normal.

(*viii*) 6th tergite of female normally concealed shorter than 5th.

(*ix*) Abdominal setulae acuminate and pale.

(*x*) 5th abdominal tergit not longer than wide.

(*xi*) Abdominal tergites 3-5 predominantly yellow to orange brown.

(*xii*) Abdominal tergites with medial dark strip on T3-T5 and a transverse dark line on T3, with medial dark stripe, usually in T3-T5, or with medial dark stripe, not brown with medial T-shaped yellow mark.

12.6. KEY TO DISTINGUISHING CHARACTERS OF MAJOR FRUIT FLIES SPECIES

After establishing the identities of different species prevalent in Kakori block of distt. Lucknow. A taxonomic pictorial key is formulated for an easy identification of different species by taking into consideration, their distinguishing characters only,

1. Chaetotaxy of head and thorax reduced (ocellar, postocellar, dorsocentral, presutural, sterno-pleural and humeral bristles absent). Third antennal segment three times longer than wide, abdominal terga not fused (seprate), wing hyaline having brown streak through cubital area (Fig. 70) .... genus *Bactrocera.*

— Abdominal tergites fused (Fig. 71) .... *Dacus*

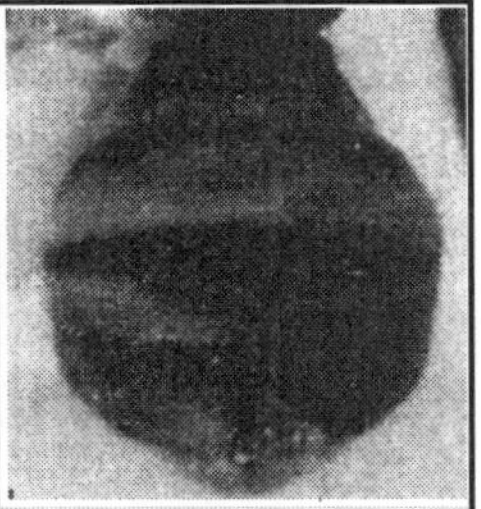

FIG. 70.

FIG. 71.

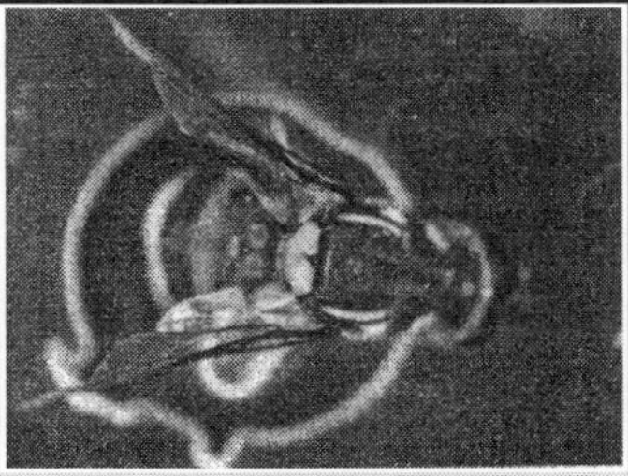

FIG. 72.

2. Vth Abdominal sternum of male with deep concavity on posterior margins, medial post sutural vitta present (Fig. 72) .... 3.

— Medial post sutural vitta absent, pecten present .... 5.

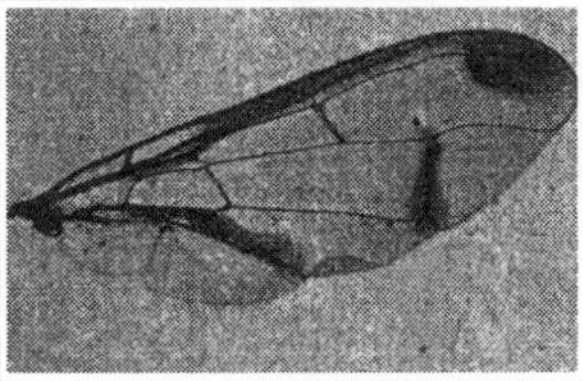

FIG. 73.

3. Pecten present scutum orange brown wings with costal band broadened into large apical spot, a prominent brown infuscation along cross vein r-m (Fig. 73), attracted to cue-lure (Fig. 74)............ ....*Bactrocera cucurbitae.*

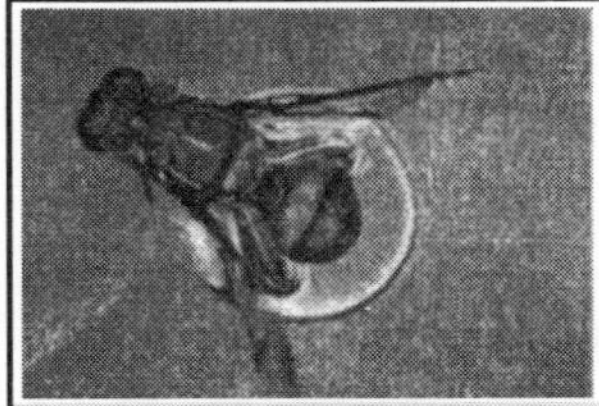

FIG. 74.

— Scutum brownish black, wing without marking along cross veins.. 4.

4. Pecten absent, dark coloured with antennal spot absent, post pronotal lobe entirely pale, mdedian vittae present, male without pecten on IIIrd abdominal tergum, attracted to methyl eugenol (Fig. 75) .... *B. diversa.*

— Pecten present, antennal spot oval .... (6)

5. All femora entirely yellow, abdominal with prominent, yellow 'T' shaped mark. Attracted to methyl eugenol (Fig. 76) .... *B. dorsalis.*

— Femora not yellow coloured .... (*Bactrocera* spp.)

FIG. 75.

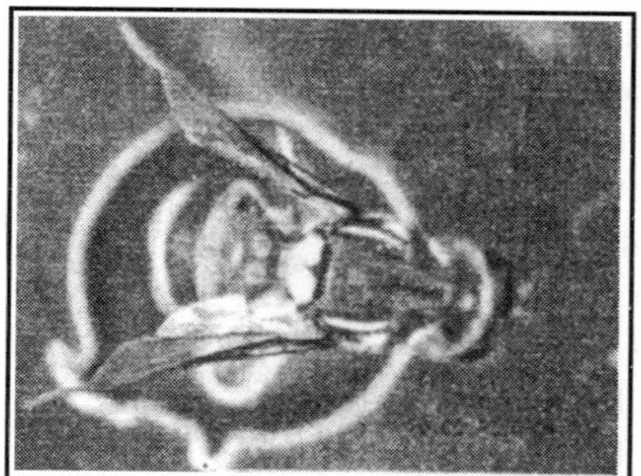

FIG. 76.

6. Scutellum orange brown, prescutellar sitae present, coastal band of wing not extended apically, face with one cross band in furrows, antennal spot oval, attracted to methyl eungenol.. *B. zonata* (Fig. 77).

— Costal band not continuous, antennal spots fused to form a transverse line, thorax black scutum yellow, cellular as long as vein abdomen refous, no cubital strease, attracted to methyl eugenol.................................... *B. correcta* (Fig. 78).

FIG. 77.

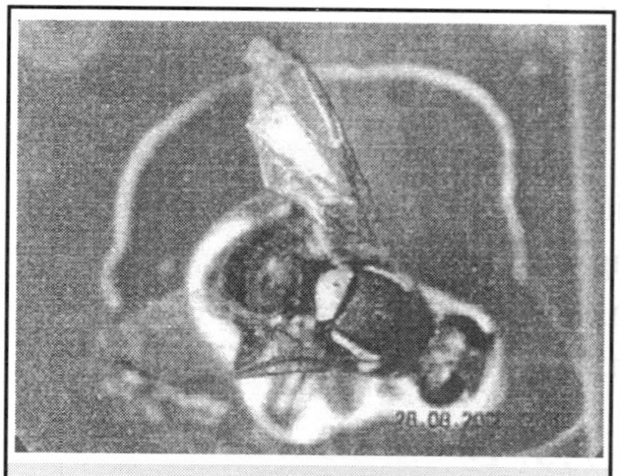

FIG. 78.

12.7 MORPHOMETRIC STUDY OF DIFFERENT FRUIT FLY SPECIES

The morphometric measurement of different body parts was taken with the help of ocular micrometer. The measurement of different body parts help in the identification of fruit flies. Shape and size of fruit flies varied in different species. The measurement of body parts such as wing span, diameter of eyeball, length of thorax, head and pretarsus were taken with the help of ocular microscope. The data recorded on 5 individuals of each species for the measurement, showed differentiation in the species of fruit fly.

TABLE 11 : Measurements of various species of fruit flies

Body parts	*Measurement of different fruit flies species (mm)*				
	B. dorsalis	*B. zonata*	*B. correcta*	*B. diversa*	*B. cucurbitae*
Body length	7.82	8.20	6.77	7.90	8.75
Body width with expanding wings	14.72	14.58	8.86	14.70	15.26
Wing Span					
(a) Length	6.82	5.77	5.24	6.72	7.17
(b) Width	3.10	2.96	2.13	3.02	3.00
Eye ball diameter	1.84	1.53	1.20	1.60	1.75
Length of thorax	2.37	2.38	1.92	2.40	2.63
Length of vittae					
(a) Lateral	1.94	2.20	1.65	1.78	2.36
(b) Medial	–	–	–	0.72	0.94
Length of abdomen	2.12	2.30	1.84	2.53	2.90
Length of pretarsus	0.11	0.21	0.10	0.12	0.12

12.7.1. *B. dorsalis*

The data recorded after morphometric analysis of *B. dorsalis* revealed that the body length of this species was 7.82 mm, width with expanded wing was 14.72 mm, wing span length 6.82 mm, wing span width 3.10 mm, length of thorax 2.37 mm, eye ball diameter 1.84 mm, length of lateral vittae 1.94 mm, length of abdomen 2.12 mm and length of pretarsus was 0.11 mm.

12.7.2. *B. zonata*

The data recorded on the *B. zonata* showed that the body length of this species of fruit fly was slightly similar to *B. dorsalis* with 8.20 mm body length, width with expended wing was recorded as 14.58 mm, length of wing span 5. 77 mm, width of wing span 2.96 mm, length of thorax 2.38 mm, eye ball diameter 1.53 mm, length of lateral vittae 2.20 mm, length of abdomen 2.30 mm and length of pretarsus was 0.21 mm.

12.7.3. *B. correcta*

The morphometric analysis of *B. correcta* showed that the body length was 6.77 mm, width with expended wing 8.86 mm, wing span length 5.24 mm, wing span width 2.13 mm, length of thorax 1.92 mm, eye ball diameter 1.20 mm, length of lateral vittae 1.65 mm, length of abdomen 1.84 mm and length of pretarsus was recorded as 0.10 mm.

12.7.4. *B. diversa*

The measurements of *B. diversa* were recorded under microscope and data showed that body length was 7.90 mm, width with expended wing was 14.70 mm, wing span length was 6.72 mm, width of wing span was recorded as 3.02 mm, length of thorax was found as 2.40 mm, eye ball diameter was 1.60 mm, lateral vittae was 1.78 mm, abdomen was 2.53 mm long and pretarsus was 0.12 mm long. According to morphometric analysis, *B. diversa* species of fruit fly was physically very similar to *B. dorsalis.*

12.7.5. *B. cucurbitae*

The body length of *B. cucurbitae* was measured, as 8.75 mm, while other body parts were measured as width with expended wing span 15.26 mm, length of wing span 7.17 mm, width of wing span 3.00 mm, eye ball diameter 1.75 mm, length of thorax 2.63 mm, length of lateral vittae 2.36 mm, length of medial vittae 0.94 mm, length of abdomen 2.90 mm and length of pretarsus was 0.12 mm.

These observations showed that the *B. cucurbitae* species was the biggest species of fruit fly. It showed highest body length, width and body width with expended wing span, while *B. correcta* was the smallest species of fruit fly.

12.8. IDENTIFICATION MARKS OF LARVAE OR IMMATURE STAGE

For the taxonomic study of immature stages of different fruit fly species, fruit flies were reared in the laboratory of Deptt. of Entomology on mango, guava and banana fruits. Fresh fruits were kept in wooden cages containing 6 cm deep layer of moist sand. 10 pairs of each fruit fly species were released for mating and oviposition. Larvae were observed under microscope for their taxonomic characters.

12.8.1. Identification marks of larvae of oriental fruit fly *B. dorsalis* Hendel

Body length 7.5-10 mm, slender, elongate and tapering anteriorly. Integument insclerotized, entirely whitish to yellowish. Caudal ridge present. Mature larvae able to jump.

Head

Head of normal shape and cephalic lobes well developed. Antenna 2-segmented.

Stomal organ

Primary lobe small round, number of peg sensilla three, or four, peg sensilla unbranched, or with few short branches, other peg like sensilla absent, sclerotized stomal guards absent. Oral ridge present, No. of oral ridge 11-14.

Caudal segments and lobes

Sensilla on caudal segment 10 pairs with atleast 7 pairs visible under disecting microscope. Ventral caudal sensilla absent or visible only with great difficulty or obvious, but not on a papilla or tubercle anal lobes plainly visible and simple.

Cephalopharyngeal skeleton

Mandibles

Subapical teeth absent, base elongate, forming a more oblique angle.

Dorsal spinules on segments T1-T3.

Anterior spiracles

Anterior spiracles elevate, margin convex to straight. Anterior spiracular tubes 8-12 in a single conform row.

Posterior spiracles

Posterior spiracular area not distinctly set off from caudal segment. Posterior spiracles, slits 2.5-3 × longer than wide. Number of dorsal spiracular processes 17-20. Number of ventral spiracular process 17-20. number of lateral spiracular re-processes 8-12.

12.8.2. Identification marks of larvae of peach fruit fly *B. zonata* Saunders

Body

Body length 9-12 mm, slender, elongate, and tapering anteriorly. Integument unsclerotized, entirely whitish to yellowish. Caudal ridge present. Mature larvae able to jump.

Head

Head of normal shape, antenna 2-segmented.

Stomal organ

Primary lobe small round, other peg like sensilla like sclerotized, stomal guards absent. Number of oral ridge 10-11 (deep, clearly defined rows margins serrate or dentate.

Caudal segments (a8) and lobes

Sensilla on caudal segment absent or visible only with great difficulty. Lobes plainly visible but not strongly protuberant simple.

Cephalopharyngeal skeleton

Mandibles

Subapical teeth absent. Parastomal bars elongate, free from hypopharyngeal sclerite. Dental sclerites present, posterior to mandibles.

Spinules and creeping welts

Dorsal spinules on segment T1-T3.

Anterior spiracles

Anterior spiracles elevate, margin concave medially, appearing bilobed. Anterior spiracular lobes 13-15, in a single uniform row or in a single irregular row.

Posterior spiracles

Posterior spiracles slits 3-3.5 times longer than wide. Number of dorsal spiracular prossesses 3-17. Number of ventral spiracular process 3-17. Number of lateral spiracular processes 6-8.

12.8.3. Identification marks of larvae of peach fruit fly *B. correcta* (Bezzi.)

Body

Body length 6.5-10 mm, slender, elongate, and tapering anteriorly. Integument unsclerotized, entirely whitish to yellowish. Caudal ridge present. Mature larvae able to jump.

Head

Head of normal shape and antenna 2-segmented. Oral ridge present, number of oral ridges 9-11 (deep clearly defined rows).

Cephalopharyngeal skeleton

Mandibles and subapical teeth absent. Dental sclerites present.

Spinules and creeping welts

Dorsal spinules on segment T1-T3.

Caudal segments (a8) and lobes

Sensilla on caudal segment 10 pairs with atleast 7 pairs visible in dissecting microscope. Ventral caudal sensilla absent.

12.8.4. Identification marks of larvae of peach fruit fly *B. diversa*

Body length 7.5-10 mm, slender, elongate and tapering anteriorly. Integument unsclerotized, entirely whitish to yellowish. Caudal ridge present. Mature larvae able to jump.

Head

Head of normal shape, antenna 2-segmented.

Stomal organ

Primary lobe small round, other peg like sensilla like structures, sclerotized stomal guards absent. Oral ridge present.

Caudal segments (a8) and lobes

Sensilla on caudal segment 10 pairs with atleast 7 pairs visible under dissecting microscope. Ventral caudal sensilla absent or visible only with great difficulty.

Cephalopharyngeal skeleton

Mandibles

Subapical teeth present. Tooth much smaller, ten apical tooth and delicate with a single ventral tooth and a single lateral tooth.

Spinules and creeping welts

Dorsal spinules on segments.

Anterior spiracles

Anterior spiracular tubules in a single uniform row on in a single irregular row.

Posterior spiracles

Posterior spiracles area not distinctly set off from caudal segment.

12.8.5. Identification marks of larvae of melon fruit fly, *Bactrocera cucurbitae* (Coquillett)

Body

Body length 9-11 mm, slender, elongate, and tapering anteriorly. Integument unsclerotized, Entirely whitish to yellowish or partially sclerotized (mature larvae), sclerotization forming a transverse line beneath posterior spiracles (mature larvae), sclerotized process on caudal segment absent. Caudal ridge present. Mature larvae able to jump.

Head

Head of normal shape. Cephalic lobes well developed. Antenna 2-segmented.

Stomal organ

Primary lobes small, round, number of peg sensilla 2 or 3 (small) and unbranched. Other peg sensilla like structures absent.

Cephalopharyngeal skelton

Mandibles

Subapical teeth present. Tooth much smaller than apical tooth and delicate (well defined).

Spinules and creeping welts

Dorsal spinules on segments T1-T3.

Caudal segments (a8) and lobes

Sensilla on caudal segment 10 pairs with atleast 7 pair visible under dissecting microscope. Ventral caudal sensilla absent or visible only with great difficulty. Anal lobes plainly visible but not strongly protuberant and simple.

Anterior spiracles

Anterior spiracles elevate, margin concave to straight. Anterior spiracular tubules 16-20 in a single uniform row.

Posterior spiracles

Posterior spiracles area not distinctly sec off from caudal segment. Posterior spiracles slits 2.5-3 times longer than wide. Number of dorsal spiracular processes 6-12. Number of ventral spiracular prossesses 6-12. Number of lateral spiracular processes 4-6. Area between posteror spiracles smooth.

❑❑❑

13
Chapter

Recent Researches on Fruit Flies

The fruit flies have been reported to be a major pest of fruits. The available literature on the trapping technique for fruit flies in guava and mango seems to be rare and limited in the country and abroad. However, detailed exhausted and up-to-date appraisal of the available literature in India and abroad has critically been done on different fruit fly species infesting fruit crops, relationship between trap catches and weather parameters and fruit fly response to different attractants in guava and mango orchards. In addition to this the species compositions of fruit flies infesting different varieties of guava and mango in laboratory has also been reviewed.

13.1. FRUIT FLY SPECIES PREVALENT IN FRUIT ORCHARDS

13.1.1. Work done in India

Guava

Makhmoor and Singh (1998) reported that *B. dorsalis* population peaked during 20 July to 3 August in guava orchards at Jammu and Kashmir, which almost coincided with fruit ripening. Highest trap catch (170.66) was obtained in the 6th week (20-27th July) when maximum and minimum temperatures were 33.6 and 25.5 degrees C, and relative humidity was 90.3 and 57%, respectively. Gupta and Bahtia (2001) studied the *Bactrocera dorsalis* and *B. zonata* population in guava orchard with the help of bottle traps containing

100 ml aqueous solution of 0.1% methyl eugenol and 0.25% malathion per trap. The corresponding catch was 427.2 and 517.0 during the 37th and 39th standard weeks. Madhura and Viraktamath (2003) recorded five species of fruit flies *viz., B. dorsalis, B. correcta, B. verbascifoliae, B. affinis* and *B. zonata* which were attracted to methyl eugenol traps located at Bangalore. Viraktamath and Sureshbabu (2004) observed that in guava, *B. dorsalis* had one peak (45th standard week) at Dharwad and three peaks (43rd, 45th and 47th standard weeks) at Kumbapur. Ravikumar (2005) observed three peaks of *B. dorsalis* during 27th, 45th and 48th standard weeks while *B. correcta* had the major peak during 27th standard week in 2004. *B. zonata* had two peaks during 11th and 18th standard weeks of 2005 in guava orchard. Rajitha and Viraktamath (2006a) observed the peak incidence of *B. dorsalis* during 30th standard week (27.16 fruit flies) and that of *B. correcta* during 30th and 45th standard weeks in 2003 in guava orchard. *B. zonata* had three peaks, during 42nd standard week in 2003 and 11th and 12th standard weeks in 2004. Satarkar *et al.* (2009) reported four fruit fly species *i.e. B. caryeae, B. zonata, B. affinis* and *B. correcta* attracted to the methyl eugenol trap in guava orchard at coastal region of Goa. Dale and Patel (2010) observed that the highest number of fruit fly population was in September and the lowest in May in the guava orchard at Sardarkrushinagar, Gujarat. The peak fly population coincided with the fruiting season of guava. The highest number of fruit flies collected from fruit fly traps baited with methyl eugenol was *B. zonata* males (3665) compared with only 357 of *B. dorsalis*. Galande *et al.* (2010) observed that four species of fruit flies, *viz., Bactrocera dorsalis* (Hendel), *Bactrocera zonata* (Saunders), *Bactrocera correcta* (Bezzi) and *Bactrocera versicolor* (Bezzi) were trapped in methyl eugenol traps in guava orchards in Pune, Maharashtra. The highest mean catch in methyl eugenol trap was that of *B. dorsalis* (46.05%), followed by *B. zonata* (30.28%), *B. correcta* (14.81%) and *B. versicolor* (8.87%). Galande and Ukey (2011) recorded seven fruit fly species *viz., B. dorsalis, B. correcta, B. versicolor* and *B. zonata* attracted to methyl eugenol trap while in cue lure three fruit flies species *i.e., B. cucurbitae, B. gavisa* and *B. tau* in the vicinity of guava orchard in Pune region. Singh and Sharma (2013) revealed that 16 traps/acre had significantly more population of captured males of *Bactrocera dorsalis* (Hendel) and *B. zonata* (Saunders) compared to 4, 8 and 12 traps/acre by using methyl eugenol based traps. Population of fruit flies started appearing in 28th SMW on guava crop and reached maximum during 39th SMW. A total of 80,663.99 males were trapped from 28th to 39th SMW. Ukey *et al.* (2013) observed that guava orchards in Ahmednagar district were infested with *Bactrocera dorsalis, B. zonata, B. correcta* and *B. verbascifoliae*. Among these, *B. dorsalis* was the most predominant species, followed by *B. zonata*.

Mango

Mann (1996) reported that *Bactrocera dorsalis* flies were observed throughout the year in methyl eugenol baited traps in a mango orchard in Ludhiana, Punjab, India.

Population counts were low in the winter months from December to February. Low catches were recorded in July. Kumar *et al.* (1997) observed that the population of *B. dorsalis* peaked during 2nd fortnight of April (453 fruit flies/trap) and May (483 fruit flies/trap) and thereafter the population declined gradually. The peak activity of pest occurred from March to June coinciding with the fruiting period of mango crop. Verghese and Sudhadevi (1998) observed a peak population of *B. dorsalis* in June followed by a minor peak in August, 1995. In 1996 the peak population was recorded in May and August coinciding with active breeding of fruit flies and another peak during September and November. Gupta and Bhatia (2001) reported that the maximum catches of *B. dorsalis* and *B. zonata* as 98.6 to 62.6 fruit flies during 30th and 27th standard weeks in 1992 and 1993 respectively in mango orchard at Himachal Pradesh. Madhura (2001) studied the seasonal activity of fruit flies in mango orchard during 1998-2000. In 1998 peak population was observed during 3rd week of May (315.14 fruit flies/week). During 1999, two peaks were observed, during 3rd week of May (282.29 fruit flies/week) and 3rd week of June (182.43 flies/trap/week). In 2000 increase in population commenced from 4th week of April and peak population was observed during 3rd week of May. Sarada *et al.* (2001b) reported that the population of fruit flies gradually increased from the first week of February, 2000 (34.0 fruit flies) to 5th week of June, 2000 (235.0 fruit flies). The peak population was observed from May to June coinciding with fruit maturity period at Chittor district of Andhra Pradesh. Dwivedi *et al.* (2003) revealed that *B. dorsalis* was first observed in April with 3 per cent infestation which gradually increased in May (8.2%), June (9.8%) and slightly declined in July (8.3%) in mango orchard at Kanpur. Sureshbabu and Viraktamath (2003) noticed that *Bactrocera dorsalis* was the dominant fruit fly among the 4 species (including *B. zonata*, *B. correcta* and *B. cucurbitae*) captured in methyl eugenol traps established in mango orchards in Kumbapur and Hulkoti, Karnataka, India. The peak catches of this fruit fly were recorded during the 21st, 23rd and 46th standard weeks. Ravikumar (2005) observed the peak incidence of *B. correcta* during first fortnight of July in mango crop. *B. dorsalis* had two major peaks during 1st week of July (27th standard week) and 3rd week of November (47th standard week) (132.67 fruit flies/trap/week) in 2004. *B. zonata* had the major peak of 37.67 fruit flies/ week during last week of May (21st standard week). Rajitha and Viraktamath (2006b) observed one peak of *B. dorsalis* (30th standard week), four peaks each of *B. correcta* (45th, 52nd, 1st and 3rd standard weeks) and *B. zonata* (42nd, 45th, 8th and 13th standard weeks) in a mango orchard at Dharwad. Rejirani *et al.* (2012) observed that, the population of *B. dorsalis* was minimum at flowering stage which increased significantly during fruit set and reached its peak in May-June, at the fruit ripening stage when monitored using bait trap and methyl eugenol trap. Singh *et al.* (2013) observed that the species trapped by using parapheromone-insecticide lure (0.1% methyl eugenol + 0.05% malathion), *Bactrocera dorsalis* Hendel was predominant with highest seasonal catch in July. The other species, *B. zonata* (Saunders) attained its peak activity in May and its population was almost half that of *B. dorsalis* in mango orchard. Nagaraj *et al.* (2014a) observed that traps charged with 0.4 ml methyl eugenol was superior in attracting highest number of *B. dorsalis* and *B. correcta*. While highest number of *B. zonata* was recorded in the traps charged with 0.6 ml methyl eugenol in mango orchard. Nagaraj *et al.* (2014c) recorded significantly highest number of *B. dorsalis* in Mallika mango orchard followed by the Mixed orchard (11.06 and 9.48 fruit flies/trap/week respectively) while highest trap catches of *B. correcta* (12.66 fruit flies/trap/week) and *B. zonata* (7.82 fruit flies/trap/week) was recorded in Banganpalli and Alphonso orchard respectively. When the total fruit flies were considered, highest number of fruit flies were trapped in Mallika (22.38 fruit flies/trap/week) orchard followed by Banganpalli (18.65 fruit flies/trap/week). While the lowest trap catches were recorded in Alphonso orchard at GKVK campus, Bangalore and Srinivaspur, Kolar, India.

Other Crops

Rai *et al.* (2008) studied the succession of fruit flies (*Bactrocera zonata* and *B. dorsalis*) in Jammu and Kashmir, India and revealed that fruit flies occurred on ber, guava, mango, citrus and phalsa, with peak populations of 450.7, 1882.2, 1967.3, 1069.3 and 85.3 adult male fruit fly/trap, respectively. At horticulture garden, Deepa *et al.* (2009) observed that peak catches of *B. zonata* was on second week of October 2006 and second week of March 2007 with the trap catch of 32.3 fruit flies/trap/week. *B. correcta* peak catch was observed on second week of December 2006 and second week of April 2007 with the trap catch of 30 and 47.6 fruit flies/trap/week, respectively. *B. dorsalis* peak mean catch of 2.0 fruit flies/trap/week in 2006, while in 2007 peak catch was 46.6 fruit flies/trap/week during fourth week of April in horticulture orchard. In Vegetable farm, *B. zonata* had peak catch on first week of October 2006 with 26.3 fruit flies/trap/week which gradually decreased with 24.3 fruit flies/trap in fourth week of January. *B. correcta* had peak catch in second week of December 2006 with 26.6 fruit flies/trap/week and fourth of April 2007 with 37.6 fruit flies/trap/week. *B. dorsalis* was recorded normally in 2006 with minimum of 1.6 fruit flies/trap and in first week of April 2007 with 37.3 fruit flies/trap/week. Thakur *et al.* (2013) revealed that two fruit fly species, *Bactrocera zonata* (Saunders) and *B. dorsalis* (Hendel) entrapped in parapheromone based attractant-insecticide (methyl eugenol-malathion) suspension and wooden block traps in peach orchard. The activity started in April but the population was low which increased thereafter and reached a peak during July (28th SW) with an average of 74.5 males/trap.

13.1.2. Work done abroad

Khattak *et al.* (1990) observed no *Bactrocera dorsalis* activity in January-March. From April onwards, the population started to build up and a peak was reached in July with 145.75 flies/week. The population declined from August-November and there was no activity in December in a pear orchard at Peshawar, Pakistan. Seewooruthun *et al.*

(2001) observed the peak activity of *B. dorsalis* during June, followed by two small peaks in July and September in mango orchard at Mauritius. The peak activity of *B. zonata* was observed during April-June. Khan and Khaliq (2003) observed that second fortnight of August and first fortnight of September showed maximum population of fruit flies trapped by pheromones in guava orchards. The population remained present throughout the fruiting season from flowering to maturity. *Bactrocera zonata* was dominant on guava with 49.62% existence, followed by *B. dorsalis* with 46.37% existence. Kawashita *et al.* (2004) reported catches of *B. correcta, B. dorsalis* and *B. zonata* in methyl eugenol while *B. cucurbitae* in cue lure in Sri Lanka. Latif and Abdullah (2005) recorded the mean number of flies captured per month ranged from 6 to 400 by methyl eugenol baited traps. Two peaks of fly population were predominant; 1st in June (217) and 2nd during September (397) in a mango orchard in Dera Ismail Khan, Pakistan. Hong and Hui (2005) reported that fruit fly, *Bactrocera dorsalis* occurred all-year-round in Yuanjiang dry-hot valley area, Yunnan, China. Its population size remained low from December to February of next year and increased steadily from March through August until it reached the peak. Afterwards, the fruit fly population declined until November. Nboyine *et al.* (2012) studied that by using traps baited with three different attractants Methyl eugenol (ME), Terpinyl acetate (TA) and Trimedlure (TML)], a total of 39,011 fruit flies were collected from mango orchards in three agro-ecological zones in Ghana. Five fly species (in two genera), *Bactrocera invadens*, *B. cucurbitae*, *Ceratitis cosyra*, *C. ditissima* and *C. capitata* were identified. The highest relative fly densities (number of flies per trap per day, F/T/D) ranged from 1.86-31.40 and 16.25-121.39 in the 2009 and 2010 seasons, respectively. Ullah *et al.* (2012) reported that two fruit fly species *Bactrocera zonata* and *Bactrocera dorsalis* were found in the Kohat district, Pakistan where *B. zonatus* was found as dominant species (94.7%) while *B. dorsalis* 5.30% was less frequent in guava orchards. Solangi *et al.* (2014) studied that species-wise seasonal weekly average fruit fly catches were 605.65 + or –66.32/10 traps for guava fruit fly. In case of *B. dorsalis*, during 1st week of November 2012, the catches were 21/10 traps which continuously decreased and reached lowest number of catches (6/10 traps) in the 4th week of February in Pakistan. This indicates that *B. zonata* population was markedly ($P<0.05$) higher than the *B. dorsalis*. Vayssieres *et al.* (2014) showed that the exotic species, *B. invadens*, was present at high levels [mean peak of 378 flies per trap per day (FTD)] in all agro-ecological zones, while the native species, *C. cosyra*, preferred the drier zones of West Africa, with lower population levels (mean peak of 77 FTD) in mango orchard. Nan *et al.* (2014) monitored population of fruit flies by using methyl eugenol trap at Yezin, Myanmar in mango, guava, and jujube fruits. Eleven species of fruit flies, *Bactrocera arecae, B. carambolae, B. correcta, B. dorsalis, B. kandiensis, B. latilineola, B. malaysiensis, B. neocognata, B. raiensis, B. verbascifoliae*, and *Carpomya vesuvina*, were identified. *B. correcta* and *B. dorsalis* were the most abundant and accounted for 29.3% and 28.6% of total emerged adults in the different fruit samples.

13.2. RELATIONSHIPS BETWEEN TRAP CATCHES AND WEATHER PARAMETERS

13.2.1. Work done in India

Guava

Gupta and Bhatia (2001) observed significant positive correlation between trap catches of *B. dorsalis* and *B. zonata* with maximum and minimum temperature. Jalaluddin *et al.* (2001) observed a significant positive correlation between the population of *B. correcta* and mean maximum temperature (r = 0.3314), minimum temperature (r = 0.3610), day-degrees (thermal units) (r = 0.3692), morning relative humidity (r = 0.4369) and rainfall (r = 0.2364) and a low negative correlation with weekly mean sunshine hours.

Sarada *et al.* (2001b) found that the fruit fly incidence had significant positive correlation with maximum temperature and non-significant and positive correlation with minimum temperature at Tirupati. Viraktamath and Sureshbabu (2004) observed that trap captures of *B. zonata* and *B. correcta* had significant positive correlation with maximum temperature and a highly significant negative correlation with morning relative humidity at Dharwad. At Kumbapur, both the species had positive correlation with maximum temperature and relative humidity. Rajitha and Viraktamath (2006a) observed that, *B. dorsalis* had significant positive correlation with minimum temperature and morning and afternoon relative humidity. It had a highly significant negative correlation with maximum temperature. Trap captures of *B. zonata* showed a highly significant positive correlation with maximum temperature. Trap catches of *B. correcta* had no significant correlation with weather parameters. Dale and Patel (2010) revealed that fruit fly population in the guava orchard had a significant positive correlation with minimum temperature and relative humidity. Singh and Sharma (2013) reported that population was positively correlated with relative humidity and sunshine but maximum temperature, minimum temperature, mean temperature, wind speed, rainfall, evaporation and number of rainy days had negative impact. The peak activity of fruit flies was found coinciding with the maturity of fruit.

Mango

Bagle and Prasad (1983) observed significant positive correlation between weekly catch of *Dacus dorsalis* Hendel and maximum temperature, while negative correlations were observed between weekly catch and relative humidity, rainfall and wind velocity in mango orchards in Karnataka, India. Agarwal *et al.* (1995) observed a significant positive correlation between trap catches of *B. dorsalis* and maximum and minimum temperature.

Kumar *et al.* (1997) reported that the population of *B. correcta* was positively correlated with maximum and minimum temperature, whereas, rainfall, sunshine hours and relative humidity had no significant effect on population variation. Agarwal and

Kumar (1999b) obtained a positive correlation between trap catches of *B. zonata* and maximum temperature, minimum temperature and rainfall and a negative correlation with relative humidity. Gupta and Bahtia (2001) revealed that there was a significant positive correlation between the trapped fruit fly (*Bactrocera dorsalis* and *B. zonata*) population and maximum and minimum temperatures in mango and guava orchard. According to Madhura (2001), the population of fruit flies showed a non-significant correlation with maximum temperature during 1998 and a significant negative correlation during 1999, and a significant positive correlation during 2000. Relative humidity had a negative correlation during 1998 and 2000, and a positive correlation during 1999. Sureshbabu and Viraktamath (2003) found that the correlation was positive and non-significant with maximum temperature. *B. zonata* and *B. correcta* had a significant positive correlation with minimum temperature at Dharwad, while at Kumbapur the species had positive correlation with maximum and minimum temperature. Rajitha and Viraktamath (2006b) observed a highly significant positive correlation between trap catches of *B. dorsalis* and minimum temperature, morning and afternoon relative humidity. *B. correcta* had a significant positive correlation with minimum temperature while *B. zonata* had a significant positive correlation with maximum temperature. Verghese *et al.* (2006) evaluated that a weak negative relationship with maximum temperature, relative humidity I and rainfall and a moderate but non-significant relationship with relative humidity II and evaporation. However, trap catches showed a significant positive correlation with minimum temperature and wind speed for the Oriental fruit fly (*Bactrocera dorsalis*) infesting mango. Munj *et al.* (2013) reported that in Vengurle, Maharashtra, India, during 2002, unseasonal rains (URs) of 98 mm were received during 16th meteorological week (MW), which increased the fruit fly (*B. dorsalis*) population (FFP) during subsequent MWs from 324 to 524 fruit flies per trap (FF/trap). Another span of URs occurred during 19th and 20th MW which enhanced the FFP from 524 to 1911 FF/trap during subsequent MWs. During 2003, a trace amount of rain was received during 12th, 14th, 17th and 20th MWs, which increased the FFP from 153.50 to 1909.25 FF/trap. During 2004, URs of 54.6, 38.2 and 21.2 mm were received during 19th, 20th and 21 MWs, which increased the FFP from 321.75 to 1508 FF/trap. During 2005, URs of 10 and 32 mm were received during 15th and 16th MWs, which increased the FFP gradually from 135.25 to 570.50 FF/trap at 19th MW. URs of 40 and 8 mm were received during 12th and 13th MWs of 2008, which increased the FFP from 34.25 to 427.50 FF/trap during 17th MW. There was a positive correlation between the URs and the FFP. The URs during the fruit growth period increased the fruit fly incidence in mango.

Other crops

Rai *et al.* (2008) reported that positive correlation between fruit fly and abiotic factors (temperature, humidity and rain) existed, whereas temperature (maximum and minimum) ranging between 15 and 30 degrees C showed highly significant impact compared to the relative humidity on population fluctuations of the fruit flies. Temperatures below 15 degrees C was lethal for the growth and development of fruit flies and during this period (December and January), fruit fly populations occurred on ber, guava, mango, citrus and phalsa were very low at Jammu and Kashmir, India. Deepa *et al.* (2009) revealed that *B. dorsalis* trap catches had positively correlated with temperature and negatively correlated with rainfall and relative humidity. Agrawal and Deepa (2013) observed that temperature had a positive correlation with population of *B. correcta* at three places *viz.* horticulture orchard, vegetable farm and IIPR while at two places it had negative correlation (r = –0.034 and 0.2811) *i.e.* at insectary and Vishnupuri, respectively at Kanpur district, (U.P.). Relative humidity was negatively correlated at four locations while it showed positive correlation at IIPR (r = 0.273). Rainfall had a positive correlation with catches of *B. correcta* at three locations while it showed positive correlation at two places. *B. zonata* and *B. dorsalis* had a positive correlation with temperature mostly.

13.2.2. Work done abroad

Hong and Hui (2005) observed that temperatures, rainfall and host fruits were the major factors influencing the population fluctuation of fruit fly, *Bactrocera dorsalis*. The monthly mean temperatures fell into the ranges of temperatures suitable for development and reproduction of the fly. But, the monthly mean minimum temperatures from December to February of next year appeared to be lower, and were suggested to be responsible for the low populations in this period. In contrast, the monthly mean maximum temperatures in May were too high and may have negative effect on the fly populations. Rainfall was another essential factor influencing the population fluctuations. Rainfall in summer months from June through August was 100-150 mm, increasing the fly population. Hong and Hui (2006) observed that male adults fly (*B. dorsalis*) had flight activities during the day time and not at night, with flying peaks at 8-9.00 and 18-20.00 h. The number of flights of the pest was lowest at 2.00 pm. The pest was more active at a light condition of 100-200 lux. Air temperature was generally suitable for the flight activity of *B. dorsalis*. Relative humidity of less than 60% adversely affected the flight activity of the fly. Elnagar *et al.* (2010) observed that *B. zonata* manifested a higher threshold of temperature than *C. capitata*. *B. zonata* appeared to prefer warmer conditions and seemed well adaptable to hot climates. The milder climate conditions in spring, autumn and early winter was favoured by *C. capitata*, while extreme summer high temperatures supported survival of *B. zonata*. Both fruit fly species disappeared during extreme cold winter on a range of fruits including mango, guava, apricot, peach, apple and pear in Egypt. Nan *et al.* (2014) observed that populations of fruit flies were positively correlative with rainfall, minimum temperature and relative humidity, and negatively correlative with the duration of sunshine in mango, guava, and jujube orchards at Yezin, Myanmar. Solangi *et al.* (2014) observed that there was positive and significant correlation between fruit fly infestation and relative humidity (r = 0.7244**), fruit fly infestation and temperature (r = 0.5299**), between

B. zonata infestation and relative humidity (r = 0.7368**), between *B. zonata* infestation and temperature (r = 0.5215**), and between *B. dorsalis* infestation and relative humidity (r = 0.4356*), suggests a simultaneous increase in fruit fly infestation with increasing relative humidity and temperature in guava at Pakistan.

13.3. FRUIT FLY RESPONSE TO DIFFERENT ATTRACTANTS IN GUAVA AND MANGO ORCHARDS

13.3.1. Work done in India

Sood and Nath (1998) found that jaggery was the best food attractant for *B. tau* (Walker) followed by the mixture of ethyl methyl ketone + ammonium acetate, ethyl methyl ketone sugar, sugar + ethyl methyl ketone + ammonium acetate + water. Kumar and Agarwal (1998) reported that mango pulp + malathion attracted maximum *B. dorsalis* at Pusa. Agarwal and Kumar (1999a) found that treatment comprising methyl eugenol, mango pulp and malathion, resulted in maximum trapping of *B. zonata* (164.52 flies/trap/week) at Pusa. Combination of protein hydrolyzate, methyl eugenol and malathion attracted 151.74 fruit flies/trap/week. Madhura (2001) observed that papaya and guava based food baits attracted significantly more *Bactrocera* spp. at Bangalore.

Sunandita and Gupta (2001) reported that bait mixtures consisting of boric acid and protein hydrolyzate caused 80 per cent mortality of *B. tau* at 10 and 12 per cent concentration of the toxicant. Dwivedi *et al.* (2002) found that all four attractants, *viz.* jaggery, gur, methyl eugenol and combination of jaggery + babul gum (*Acacia nilotica*) were good in trapping. Each attractant was mixed with 0.05% malathion and poured into the troughs. The maximum number of fruit fly (*Bactrocera dorsalis*) was trapped by methyl eugenol followed by jaggery. Poor performance was observed in the gur attractant at a guava orchard in and around Allahabad, Uttar Pradesh, India. Rajitha and Viraktamath (2005) tested various protein baits, among them yeast, soybean, fruit fly diet, protein and casein were more female selective, when total fruit flies were considered. Soybean + sugar + banana was the most superior protein bait with a fruit fly capture of 4.5/trap/week in guava, while casein + sugar + papaya attracted more female fruit flies with a mean capture of 4.33 in mango. Ravikumar (2005) observed that fruit fly diet and casein based protein baits were promising against *B. dorsalis* in mango and guava orchard while *B. cucurbitae* was attracted to fruit fly diet, soybean, yeast in the same orchards. Kumar and Viraktamath (2007) evaluated the efficacy of different protein sources mixed with sugar and banana pulp (1 : 1 : 1) and ammonium acetate (5%) in attracting female adults of fruit flies. In guava orchard, *Bactrocera correcta* (Bezzi) responded poorly to all the baits. However, *B. dorsalis* (Hendel) females were attracted to bait containing casein or commercial fruit fly bait in significant numbers (3.25 and 3.5/trap/week, respectively). The adults of *B. cucurbitae* (Coq.) showed significantly higher preference to baits containing commercial fruit fly diet and yeast (4.0 and 3.25/trap/week, respectively).

Irrespective of species, bait containing commercial fruit fly diet, casein and yeast were more attractive to females (8.0, 7.0 and 5.75/trap/week, respectively). A similar trend was evident in the mango orchards at Dharwad, Karnataka.

Ravikumar and Viraktamath (2007) reported that food bait containing proteinex and 5 per cent ammonium acetate attracted significantly more *Bactrocera correcta* (5.17 fruit flies/trap/week), *B. dorsalis* (9.42 fruit flies/trap/week), *B. cucurbitae* (2.25 fruit flies/trap/week) and total fruit flies (16.84 fruit flies/trap/week) in guava. While in mango, fruit fly diet and mango pulp combined with 5 per cent ammonium acetate were attractive to *B. dorsalis* (7.63 and 4.63 fruit flies/trap/week, respectively), *B. cucurbitae* (3.00 and 4.25 fruit flies/trap/week, respectively) and total fruit flies (10.63 and 8.88 total fruit flies/trap/week, respectively). Food baits containing guava pulp with ammonium acetate in guava (5.59 total fruit flies/trap/week) and casein with ammonium acetate, mango pulp and proteinex with 5 per cent acetic acid (3.18, 4.63 and 3.51 total fruit flies/trap/week, respectively) were the next best treatments in mango at farmers' commercial guava and mango orchards near Dharwad. Nagaraj *et al.* (2014b) studied that casein + sugar + mango and casein + sugar + banana was the most superior food bait attracting significantly highest number of male and female fruit flies with a mean trap catch of 26.00 and 7.00 fruit flies/trap/week respectively. Traps baited with casein + sugar + mango was the most superior food bait attracting significantly highest number of fruit flies with a mean trap catch of 31.66 fruit flies/trap/week.

13.3.2. Work done abroad

Jiron and Hedstrom (1991) reported that a total of 10 446 fruit flies were trapped in traps baited with Torula yeast plus borax as an attractant in mango and guava orchards in Costa Rica. Wakabayashi and Cunningham (1991) found that 4-component mixture, made up of ammonium bicarbonate, linolenic acid, putrescine and pyrrolidine, were effective food bait for *Dacus cucurbitae* (*Bactrocera cucurbitae*). Alyokhin *et al.* (2000) observed that baiting sticky Ladd traps with hydrolysed liquid protein significantly increased the number of captured oriental fruit flies, *Bactrocera dorsalis* in an orchard of commercial guava, in Hawaii, USA. Mostly male flies were caught in the absence of mature guava fruit, whereas mostly female flies were caught when ripe fruit was abundant. Corneluis *et al.* (2000a, b) reported that liquid hydrolyzed proteinaceous bait (Nulure) attracted more female oriental fruit flies in guava orchard as compared to several ammonia based olfactory lures. Sewoosunkur *et al.* (2000) observed that the locally-made bait was generally as effective when used in McPhail traps as commercially obtained protein hydrolysate bait. Fruit fly catches were greater when protein autolysate was prepared with higher concentrations of papain (2.0 or 4.0 g/litre). Thomas *et al.* (2001) observed that open bottom plastic traps baited with two component synthetic lure (ammonium acetate and putrescine) caught more fruit flies than McPhail traps baited with Torula yeast. Fabre *et al.* (2003) reported that solbait (protein hydrolyzate) was the most effective in capturing females of melon fly. Ho *et al.* (2003) revealed that poisoned protein hydrolysate bait in the improved McPhail trap in guava orchards caught 18.1 and 19.5% of females and males, respectively. Duyck *et al.* (2004) revealed that adding borax to protein hydrolysates Nulure and Buminal strongly reduced their attractiveness for *B. cucurbitae*. Acidification of 5% Buminal solution (from pH 6 to pH 3) doubled its attractiveness for melon fly. Conversely, Torula yeast at pH 10.5 was significantly more attractive than the standard Torula yeast at pH 9 (28% of captured flies compared with 17%). Katsoyannos and Papadopoulos (2004) reported that spheres baited with three component food attractant containing ammonium acetate, putrescence and trimethyl amine was most attractive to female *C. capitata*. Braga Sobrinho *et al.* (2004) observed that the highest captures of female adults of *Ceratitis capitata* in mango and guava orchards were obtained from combinations of ammonium acetate (AA) + putrescine (PT) + trimethylamine (TMA) + H2O and AA + PT + TMA + propylene glycol (PG). The highest adult captures (5397 males and 5962 females) of *Anastrepha* spp. in guava orchards came from traps baited with nulure (NL) + H2O. The best trap efficiencies for female adults of *C. capitata* were obtained by combinations of NL + H_2O, AA + PT + TMA + Triton and AA + PT + TMA + PG. However, for *Anastrepha* spp. the best efficiencies came with NL. In mango orchards, the best combination for *C. capitata* was AA + PT + TMA and for *Anastrepha* spp. was AA + PT + TMA + PG. The treatment NL + H_2O allowed the highest fly per trap per day (FTD) indices for both fruit flies in Ceara, Brazil. Latif and Abdullah (2005) revealed that significant reduction in fruit infestation when traps were adjoined either by applying bait of protein hydrolysate or neem oil separately and in combination in a mango orchard at Dera Ismail Khan, Pakistan. Seewooruthun *et al.* (2005) reported that the synthetic food attractants based on ammonium acetate alone or in combination with other attractants, appear to be a more effective and selective option for fruit fly surveillance including *B. zonata*, *B. cucurbitae*, *Ceratitis rosa* and *C. capitata*, than the more conventional hydrolysate protein NuLure and the Torula yeast. Medeiros *et al.* (2011) concluded that the food attractive based on mango and guava juice had the best attraction in the county of Bananeiras-PB. In the assays, species of fruit flies *viz.*, *Ceratitis capitata, Anastrepha fraterculus, A. sororcula, A. zenildae* and *A. obliqua* was found. Vayssieres *et al.* (2011) reported that traps contained with three attractants *viz.* Terpinyl acetate, Trimedlure and Torula captured 77642 fruit flies in more than 4 months including mango season; they were represented by 18 identified fly species, including ten *Ceratitis*, six *Dacus* and two *Bactrocera* species in mango orchards of Niayes region, Senegal. Mahmoud *et al.* (2012) found that Torula yeast was attractive for eight consecutive weeks when appropriate amount of water added to traps to compensate the evaporated amount. Ciproflaxocine significantly enhanced the attraction capacity of Nulure and Mazoferm among tested attractants to fruit flies that infest guava and mango in Sudan. Ali *et al.* (2013) reported that there was a significant difference between Lynnfield trap baited with methyl eugenol (42.3) and Lynnfield trap baited with local bait protein hydrolysate (0.6) for capturing *B. invadens, C. cosyra and C. capitata* in mango and guava. Lynnfiled trap baited with methyl eugenol was more efficient and attracts large number of fruit flies than a local bait trap.

El-Gendy (2013) observed that Ammonium Acetate (AA) was the highest potential attractants in attracting adults of Peach Fruit Fly (PFF), *Bactrocera zonata*. The borax mixed with Ammonium Di-hydrogen orthophosphate (AD) enhanced the AD attraction potential for PFF compared to attraction of AD alone. The mean number of captured flies of PFF was inversely to tested time, *i.e.,* 0.23 + or –0.02 and 0.16 + or –0.03 capture flies/trap/day in the first and second tested weeks. Toth *et al.* (2014) reported that the addition of the synthetic *Rhagoletis* lure (consisting of ammonium carbonate and ammonium acetate) invariably increased catches of the European cherry fruit fly (*R. cerasi*), eastern cherry fruit fly (*R. cingulata* Loew.) and walnut husk fly (*R. completa* Cresson).

13.4. SPECIES COMPOSITIONS OF FRUIT FLIES INFESTING DIFFERENT VARIETIES OF GUAVA AND MANGO IN LABORATORY

13.4.1. Work done in India

Kalia (1992) reported that *Dacus dorsalis* (*Bactrocera dorsalis*) preferred the 'Bangalora' and 'Mallika' varieties of mango followed by 'Dashehari' and 'Amrapali'. The survival percentage was highest on 'Bangalora' and lowest on 'Amrapali'. *B. dorsalis* preferred the guava varieties 'Lucknow 49' to 'Allahabad Safeda'. The duration of the incubation, larval and pupal periods was shortest on 'Lucknow 49' followed by 'Allahabad Safeda'. Mann (1996) reported that fruit fly, *Bactrocera dorsalis* infestation was 30.77, 65 and 85.50% in cultivars Dushehari, Sucking and Chausa, respectively, in a mango orchard in Ludhiana. Verghese *et al.* (2002) observed that the loss to mango in an unsprayed situation varied from 2.5 to 59.0% depending on the variety. The cultivars Dushehari and Langra were least infested, while cultivars Banganpalli and Totapuri had a mean infestation of 46.0 and 59.0%, respectively, in Bangalore. Singh (2008) observed that amongst seven guava varieties *viz.* Lucknow-49, Seed Less, Behat Coconut, Red Flesh, Allahabad Safeda, Local and Pear Shaped, smooth skinned guava varieties namely Red Flesh, Allahabad Safeda and Local were found to be highly susceptible to the fly, *Bactrocera dorsalis* attack (Infestation range 64.2 to 80.4%) whereas rough skinned Pear Shaped variety was least susceptible (35.1%) as compared to other varieties where the infestation ranged between 45.7 to 56.5 per cent. Nandre and Shukla (2013) investigated the reaction of sapota varieties for their resistance/susceptibility to fruit fly, *Bactrocera dorsalis*. The genotypes *viz.,* PKM-1, PKM-2, DHS-1, DHS-2, Bhuripatti, Pilipatti and Singapore were categorized as least susceptible to *B. dorsalis*. However, Zumakhiya, CO-2 and Kirthibarthi were found to be moderately susceptible. Likewise Kalipatti, Cricket

Ball, Paria Collection, Murabba and Mohangoote were recorded as highly susceptible to *B. dorsalis* in Gujarat under field and laboratory conditions. Nagaraj *et al.* (2014c) observed that highest emergence of maggot in different mango varieties was in cv. Mallika (1.40 maggot/fruit) followed by Amrapali (1.00 maggot/fruit) in laboratory study.

13.4.2. Work done abroad

Ambele *et al.* (2012) studied the susceptibility of four economically important mango varieties (Kent, Keith, Palmer and Haden) in Ghana to the attack by *Bactrocera invadens* in laboratory experiment. Kent was found most susceptible, followed by Palmer, Haden and Keith. Flies took significantly longer periods to complete development on the least susceptible variety (Keith) (25.53 ± 2.3 days), than on the most susceptible variety (Kent) (19 ± 2.3 days). Magid *et al.* (2012) revealed that among the mango varieties, the late mango varieties, Shendi and Abusamaka were more susceptible to fruit fly infestation than the early mango varieties, Kitchener and Baladi. Within the fruiting stage of maturity, the ripening stage was found to be highly infested and was the most susceptible stage for flies' infestation at Shendi area, River Nile State, Sudan. Jose *et al.* (2013) reported that four fruit fly species emerged from the collected fruits : *Bactrocera invadens*, *Ceratitis rosa*, *Ceratitis cosyra* and *Ceratitis capitata*. *Bactrocera invadens* was the most abundant species (97%), followed by *C. rosa* (1.8%), *C. cosyra* (1.1%) and *C. capitata* (0.1%). The highest damage was observed in guava (92.5% of fruits) followed by tropical almond (67.3%) and mango (56.5%) in Mozambique.

13.5. MASS TRAPPING

13.5.1. Work done in India

Ramani (1997) listed about 28 species that are predominant in India. About four species *viz., Bactrocera dorsalis, Bactrocera zonata, B. correcta* and *B. caryeae* have been reported from *Bactrocera dorsalis* complex having common character in all.

Application of insecticides further disrupt the ecosystem and cause numerous hazards, which in the present scenario warrants the need of integrated approach for fruit fly management (Verghese *et al.,* 2002, Sharma *et al.,* 2011, Verghese *et al.,* 2012). Among the various alternate strategies available for the management of fruit flies, the use of methyl eugenol traps and cue lure is also used for trapping the fruit flies. Methyl eugenol traps stand as the most outstanding alternative. Methyl eugenol has both olfactory as well as phagostimulatory action and is known to attract fruit flies from a distance of 800 meter (Roomi *et al.,* 1993). Methyl eugenol, when used together with insecticide impregnated into a suitable substrate, forms the basis of male annihilation technique. This technique has been successfully used for the eradication and control of several *Bactrocera spp.* Hancock *et al.,* 2000, Alloved *et al.,* 2003, Manzar and Srivastava 2004, Stoenhouse *et al.* 2005 and Stoenhouse *et al.,* 2007, reported that MAT with methyl eugenol traps @ 4 trap/acre in mango and guava was found effective in controlling fruit flies in different parts of India. Methyl eugenol specially attracts the male of *B. dorsalis, B. correcta, B. zonata* (Verghes *et al.,* 2006), while cue lure attracts the *B. cucurbitae, B. correcta, B. zonata,* and *B. diversa etc.*

Thus, the tendency of *Bactrocera* males to re-visit a methyl eugenol source following initial exposure was explored to study the population buildup of different fruit flies. The sanitation combined with the use of lures and traps as well as baits proved to be one of the best alternatives for management of fruit flies. These traps have high efficiency, low cost and are environmentally quite safe (Sureshbabu and Viraktamath, 2003).

Food bait containing Proteinex and 5 per cent ammonium acetate attracted significantly more *Bactrocera correcta* (5.17 fruit flies/trap/week), *B. cucurbitae* (2.25 fruit flies/trap/week) and total fruit flies (16.84 fruit flies/trap/week) in guava. While in mango, fruit fly diet and mango pulp combined with 5 per cent ammonium acetate were attractive to *B. dorsalis* (7.63 and 4.63 fruit flies/trap/week, respectively), *B. cucurbitae* (3.00 and 4.25 fruit flies/trap/week, respectively) and total fruit flies (1.63 and 8.88 total fruit flies/trap/week, respectively) (Ravikumar and Viraktamath, 2007). To control and cost-effectiveness of male annihilation technique (MAT), plywood blocks (50 mm × 50 mm × 12 mm) soaked in ethyl alcohol : methyl eugenol : malathion in the ratio of 6 : 4 : 1 for 48 hr are hung in branches @ 10/ha in bottle traps in 2 rounds together with bait application technique (BAT) (2 g carbaryl and 100g jaggery in 1 liter water), supplemented with field sanitation was compared with the insecticide spray in the mango orchard. MAT and BAT together utilized 11.11 times less insecticide and 1.32 times less cost but resulted in nearly double protection (93-97%), as well as double cost : benefit ratio (1 : 21 : 76) than that of cover spray (53-56% and 1 : 10.9, respectively). Jaggery was found better food bait, whereas plywood block was observed to be better carrier of methyl eugenol for management of fruit fly (*Bactrocera dorsalis* Hendel) in mango. (Singh *et al.*, 2008). According to Panday *et al.* (2012), NSKE + banana poison bait recorded the lowest yield loss (2.24%) due to the fruit fly damage in bitter gourd crop followed by nimbicidine + banana poison bait trap (3.16%). (Reji *et al.,* 2012) revealed that the percent infestation of Oriental fruit fly *Bactrocera dorsalis* (Handel) was significantly reduced with methyl eugenol trap (1.13-1.34%) which was on par with the chemical control in mango crop. Agrawal and Deepa (2013) caught three major species of *Bactrocera dorsalis* complex through methyl eugenol traps. Singh *et al.,* 2013 reported that by continuous trapping of fruit (*Bactrocera* spp. males), fruit fly catches declined significantly by using parapheromone-insecticide lure (0.1% methyl eugenol + 0.05% malathion) traps @ 10 traps/ha. Nagaraj *et al.,* 2014a Studied that casein + sugar) + mango was the most superior food bait attracting significantly highest number of male fruit flies over an exposure of four weeks accounting for a mean trap catch of 26.00 fruit flies/trap/week. The food baits containing casein + sugar + banana attracted significantly highest number of female fruit flies over an exposure of four weeks accounting for a mean trap catch of 7.00 fruit flies/trap/week. When the total fruit flies irrespective of species were considered, traps baited with casein + sugar + mango was the most suitable food bait attracting significantly highest number of fruit flies with a mean trap catch of 31.66 fruit flies/trap/week. The traps baited with casein + sugar + pineapple recorded lowest number of fruit flies in mango orchard. (Nagaraj *et al.,* 2014b) observed that traps charged with 0.4 ml methyl were superior in attracting highest number of *B. dorsalis* and *B. correcta.* While highest number of *B. zonata* was recorded in the traps charged with 0.6 ml methyl eugenol in mango orchard.

12.5.2. Researches done in other countries

Maximum control of Oriental fruit fly (*Bactrocera zonata*) attacks on guava was achieved in Pakistan by Male annihilation technique followed by Bait application technique (Hussain *et al*., 2010). The attractiveness of protein bait to *Bactrocera dorsalis* (Hendel) can be significantly increased by botanical attractants. A bioassay of plant extracts, hosts fruit and other plant secondary substances was 11.11% and 10.56%, respectively. With regard to plant secondary substances, the average attractant rate of methyl eygenol was 45.00%, with rates for males and females of 86.67% and 3.33%, respectively. Ethyl acetate and guava flavor were less attractive than methyl eugenol, with an average attractant rate of 32.50% and 28.33%, respectively (Wang *et al.,* 2012). Torula yeast significantly attracted the highest number of fruit flies infesting guava and mango in Sudan compared to Nulure, GF-120 and AFFI bait. Torula yeast was found attractive when appropriate amount of water added to traps to compensate the evaporated amount. Ciproflaxocine significantly enhanced the attraction capacity of Nulure and Mazoferm (Mahmoud *et al.,* 2012). There was a significant difference between Lynnfield trap baited with methyl eugenol (42.3) and Lynnfield trap baited with local bait protein hydrolysate (0.6) for capturing *B. invadens, C. cosyra* and *C. capitata* in mango amd guava. Lynnfield trap baited with methyl eugenol was more efficient and attracts large number of fruit than a local bait trap. Methyl eugenol proved to be more effective in attracting the male of *Bactrocera invadens* (Ali *et al.,* 2013). Dilution of methyl eugenol with paraffin oil reduced the quantity of methyl eugenol. Methyl eugenol-fentrithion mixture at 100% of methyl eugenol showed that the weekly mean number of captured males were significantly higher (87.3 individuals/block) then more attracted to methyl eugenol-spinosad mixture compared with methyl eugenol-(fentrithion and thiamethoxam + abamactin) mixtures (Ghanim, 2013).

The MLT (Multi lure trap) with yellow base, baited with hydrolysed protein attracted higher number of *B. oleal* males, while Torula yeast attracted higher number of female flies in Spain.

13.6. EFFECT OF COLOURED TRAPS ON FRUIT FLY CAPTURE

13.6.1. Work done in India

Rajitha and Viraktamath (2005) evaluated response of fruit flies to the traps with different size, shape and colour in mango orchard. Medium and big traps attracted significantly more flies than small traps. Fruit flies showed greater response to spheres than to the bottles and cylinders while responses to different colours varied among different species. *Bactrocera dorsalis* was more attracted to green medium and big spheres and all size orange spheres (0.45 to 0.49 fruit flies/trap/day).

Saputra and Marmaini (2016) experimented in a completely randomized design with 4 treatments and 9 replications. The treatments tested were steiner traps baited in methyl eugenol without colour treatment (control) and steiner traps baited with methyl eugenol treatment of the red, yellow and green. The results obtained revealed that there is one type of fruit flies which were trapped namely *Bactrocera dorsalis* Hendel. From the experiments, there were more fruit flies trapped in yellow colour baited with Methyl eugenol, the average number of the fruit fly trapped was (11.74), followed by the Green colour trap by the number of average (8.67), then the trap without colour (control) (7.46) and the Red traps (7.28). Toorani *et al.*, (2017) reported that Mediterranean fruit fly, *Ceratitis capitata* (Wiedemann, 1824) (Dip. : Tephritidae) is one of the most destructive agricultural pest in the world. To replace chemical control methods, as well as the timing of the peak population of Mediterranean fruit fly, the yellow sticky vertical traps were used in this study during the four months from August to November 2016. To determine the best type of yellow color to trap this fly, four types of yellow colors with measured wavelength of reflection, including the fluorescent yellow (538 nm), lemon yellow (574 nm), yellow (602 nm) and amber yellow (641 nm), were used. The results of mean comparison of captured flies in different treatments indicated that the fluorescent yellow, heights of 1.5 and 2 m, the south direction, and month October had the highest number of capture in yellow cards.

13.6.2. Work done abroad

Sivinski, (1990) experimented with coloured spheres of five different sizes and colours. They were coated with a trapping compound and presented to field caged flies. Most of the females were captured on the 20 cm orange, green, white balls and 8.9 cm yellow ball. There were no statistically significant preferences on the part of males for any particular size or colour.

Robacker (1992) reported that relative attractiveness of yellow colour compared to green was less affected by season. Overall, vertical rectangles were more attractive than spheres in spring while spheres were more attractive in autumn. In spring, red rectangles were more attractive than yellow and white spheres.

❑❑❑

Summary

Fruits are very valuable commodity whether they are consumed locally or sent to central markets or exporter. The export potential of fruits from India is largely threatened by fruit flies and other insects which cause imperfections of any kind, physical or physiological or spoilage in the package. Through a better understanding of orchard ecosystems and the dynamics of pests and their natural enemies, together with efficient monitoring of the population, chemical intervention can be minimized.

"True fruit flies" also known as Tephritid flies, belong to the Phylum : Arthopoda, Class : Insecta Order : Diptera and Family : Tephritidae. India has nearly 200 species of fruit flies, which is about 5% of the total world tephritid flies. The tribe Dacini with the genus *Bactrocera* is of importance in India, which includes *Bactrocera dorsalis* (Hendel), Bactrocera *zonata* (Saunders), *Bactrocera correcta* (Bezzi) and *Bactrocera cucurbitae* (Coquillet). *Bactrocera* spp. except *B.. cucurbitae* are pests of major importance to fruit crops which includes mango, citrus, guava, ber etc. Females insert their eggs beneath the skin of fruit, especially in ripening fruit. White banana shaped eggs are deposited in clusters. The larvae tunnel into the fruit contaminating it with frass and providing entry for fungi and bacteria. When the infested fruit is immature, it ripens prematurely and is unfit for marketing. Warm and humid weather is considered to be favorable for *Bactrocera* fruit flies and pest populations build up as ripening occurs. Populations decrease during dry periods.

The current trend in fruit fly control is towards a more coordinated fruit fly action programme, the major objective of which will be to overcome the often ineffective and unsustainable control resulting from uncoordinated approach by individual farmers. Since fruit flies attack different crops in the same area, their management must be based not only on mango/guava/vegetables but also on wild hosts and commercially grown host plants. Thus, to improve the efficiency of fruit fly management, host plant blooming and fruiting factors need to be elucidated. When fruit growers pursue a concerted fruit fly population management strategy over a significantly large area, the number of fruit flies moving into orchards from neighboring orchards is greatly reduced. Further fruit fly management programme is based on border trapping, enhancement of host plant growth regulations, application of sterile insect techniques and parasitoid augmentation releases. The success of these goods are dependedent on grower status (rich v/s poor), access to technology, cost scale (single orchard, regional level national level), globalization of markets and impact on the environment.

TABLE 12 : Infestation Cycle of Important Fruit Flies

Fruit Flies	*April-June*	*July-Aug.*	*Sept.-Oct.*	*Nov.-Dec.*
B. cucurbitae	In hot and dry places oviposition in cucurbits often mixed with *clilatus*	Peak activity period : Servere infestation on bitter guard, ribbed gourd, squash melon and long melon	Severe infestation continuous in bottle gourd, snake gourd etc.	Attack continuous in bitter gourd, sponge gourd, pumpkin and ivy gourd whichever is grown up to December. Some time even capsicum and tomatoes.
B. dorsalis complex	Oviposition during spring. In the absence of preferred hosts may attack loquat, apricot and plum; infestation substaintially higher in peaches in June.	Population at its peak infestating mango, guava, pear (Least preferred and only in the absence to preferred host); infestation in guava, often mixed with *zonata*. Mango most preferred host (so also called "Mango fly") peak attack in August-Sept. when duration of life cycle is short (about 2 days)		Low infestation continuous in winter on guava, often mixed with *zonata*.

References

Abdullah, K., Akram, M. and Alizai, A.A. (2002). Non traditional control of fruit flies in guava orchards in D. I. Khan. *Pakisthan Journal of Agricultural Research,* **17** (2) : 71-74.

Agarwal, M.L. and Kapoor, V.C. (1989). Indian Dacini (Diptera : Tephritidae) and their host plants relationship. *Bullatin of Entomology.* **30** (2) : 186-194.

Agarwal, M.L. and Kumar, P. (1999a). Relative efficacy of bait and attractant combinations against peach fruit fly, *Bactrocera zonata* (Saunders). *Pestology,* **23** : 23-26.

Agarwal, M.L. and Kumar, P. (1999b). Effect of weather parameters on population dynamics of peach fruit fly, *Bactrocera zonata* (Saunders). *Entomon,* **24** : 81-84.

Agarwal, M.L., Rahman, S., Singh, S.P.N. and Yazdani, S.S. (1995). Weather conditions and population dynamics of *Bactrocera dorsalis. Journal of Research,* Birsa Agricultural University, **7** : 149-151

Agrawal, N. and Deepa, M. (2013). Population dynamics of fruit fly species caught through methyl eugenol traps at different locations of Kanpur, Central U.P. *Journal of Entomological Research,* **37** (1) : 87-90.

Agrawal, N. and Srivastava, R. (2008). Monitoring fruit fly (*Bactrocera* species) population infesting fruit crops through male trapping technique. *Indian Journal of Environmental Sciences,* **12** (2) : 141-145.

Agrawal, Neerja and Kumar, B.R. (2008). Biology of *Bactrocera cucurbitae* Coq. and its viability on different hosts. *Insect Environment* **14** (3) : 106-107.

Agrawal, Neerja and Mathur Y.K. (1988). The fruit fly problem associated with cultivated crops in India and its control. Fruit flies in the tropics eds. Vigaysegarm. Pub. MARDI, Kuala Lumpur, Malaysia pp 140-151.

Agrawal, Neerja and Mishra, J.K. (2010). Morphometric studies of different fruit fly species belonging to genus *Bactrocera* (Tephritidae : Diptera). *Insect Environment* **15**(4) : 149-150.

Ali, S.A.I., Mahmoud, M.E.E. Wang ManQun and Mandiana, D.M. (2013). Survey and monitoring of some Tephritidae of fruit trees and their host range in River Nile State, Sudan. *Persian Gulf Crop Protection,* **2** (3) : 32-39.

Alyokhin, A.V., Messing, R.H. and Duan, J.J. (2000). Visual and olfactory stimuli and fruit maturity affect trap captures of oriental fruit flies (Diptera : Tephritidae). *Journal of Economic Entomology,* **93** (3) : 644-649.

Ambele, F.C., Billah, M.K., Afreh-Nuamah, K. and Obeng-Ofori, D. (2012). Susceptibility of four mango varieties to the Africa Invader fly, *Bactrocera invadens* Drew, Tsuruta and White (Diptera : Tephritidae) in Ghana. *Journal of Applied Biosciences,* **49** : 3425-3434.

***Anjum, S., Razaq, M. and Yazdani, S.S. (2000).** Studies on seasonal activity and control of fruit flies (*Dacus* spp.) on mango (*Mangifera indica* L.) at Faisalabad, Pakistan. *Arab Journal of Plant Protection,* **18** : 121-123.

Anonymous (2013). Indian horticulture production at a glance. **In** : N.C. Mistry, Brajendra Singh and C.P. Gandhi (Eds.), *Indian Horticulture Database 2013.* Pub. National Horticulture Board, Ministry of Agriculture, Government of India 85, Institutional Area, Sector-18, Gurgaon, India, pp. 2-5.

Anonymous (2016). Indian Horticulture database. National Horticulture Board, Horticulture Statistics Division, Department of agriculture, Cooperation and Farmers welfare, Ministry of Agriculture, Govt.of India.

Anonymous (2017). Indian Horticulture database, National Horticulture Board, Ministry of agriculture, Govt of India.

Armstrong, J.W. and Jang, E.B. (1997). An overview of present and future fruit fly research in Hawaii and U.S. mainland. In : A. J. Allwood and R.A. Drew (Eds.), *Management of Fruit Flies in the Pacific.* ACIAR Proc. No. 76. Austral. Center Int. Agric. Res., Canberra, pp. 30-42.

Armstrong, J.W.; Hu, B.K.S.; Brown, S.A. (1995). Single-temperature forced hot-air quarantine treatment to control fruit flies (Diptera : Tephritidae) in papaya. *Journal of Economic Entomology* **88** : 678-682.

Babu, K.S. and Viraktamath, S. (2003a). Population dynamics of fruit flies (Diptera : Tephritidae) on cucurbits in north Karnataka. *Pest Management and Economic Zoology* **11** (1) : 53-57.

Babu, K.S. and Viraktamath, S. (2003b). Species diversity and population dynamics of fruit flies (Diptera : Tephritidae) on Mango in northern Karnataka. *Pest Management and Economic Zoology* **11**(2) : 103-110.

Bagle, B.G. and Prasad, V.G. (1983). Effect of weather parameters on population dynamics of Oriental fruit fly, *Dacus dorsalis* Hendel. *Journal of Entomological Research,* **7**(2) : 95-98.

Balock, J.W. and Lopez, D.F. (1969). Trapping for control of Mexican fruit fly in Mango and citrus growers. *Journal of Economic Entomology* **62** : 53-56.

Bateman, M.A. (1972). The ecology of fruit flies. *Annual Review of Entomology* **17** : 493-518.

Bateman, M.A. (1982). Chemical methods for suppression or eradication of fruit fly populations. In : *Economic fruit flies of the South Pacific Region* (Ed. by Drew, R.A.I.; Hooper, G.H.S.; Bateman, M.A.) (2nd edition), pp. 115-128. Queensland Department of Primary Industries, Brisbane, Australia.

Bess, H.A., Bosch, van den and Haramoto, F.H. (1961). Fruit fly parasites and their activities in Hawaii. *Proceedings Hawaiian Entomological Society* **17** (3) : 367-378.

Batra, H.N. (1988). Biology and bionomics of Dacus (Zeugodacus) hageni de Meijere (=D. caudatus Fabr.). *Indian Journal of Agricultural Science,* **38**: 1015-1120.

Bharathi, T.E., Sathiyanandam, V.K.R. and David, P.M.M. (2004). Attractiveness of some food baits to the melon fruit fly, *Bactrocera cucurbitae* (Coquillett) (Diptera : Tephritidae). *International Journal of Tropical Insect Science,* **24** (2) : 125-134.

Bhatia, H.L. (1939). New record of two fruit flies, *Dacus ciliatus* and *Craspedoxantha octopunctata,* from Delhi. Indian journal of Entomology **1** : 108.

Bhatia, S.K. and Mahto, Y. (1968). Note on breeding of fruit flies *Dacus ciliatus* Loew and *D. cucurbitae* Coquillett in stem galls of *Coccinia indica* W. and A. *Indian Journal of Entomology* **30** : 244-245.

Braga Sobrinho, R., Mesquita, A.L.M., Enkerlin, W., Guimaraes, J.A., Bandeira, C.T. and Peixoto, M.J.A. (2004). Evaluation of fruit fly attractants in the State of Ceara-Brazil. *Revista Ciencia Agronomica*, **35** (Numero Especial) : 253-258.

Burikam, I.; Sarnthoy, P.; Charensom, K.; Kanno, T.; Homma, H. (1992). Cold temperature treatment for mangosteens infested with the oriental fruit fly. *Journal of Economic Entomology* **85** : 2298-2301.

Chiu H.T. and Chu Y.T. (1986). The occurrence and injury of the oriental fruit fly in the Southern Taiwan. Pl. Prot. *Bulletin Taiwan*; **28** : 313-32.

Christenson, L.D.; Foote, R.H. (1960). Biology of fruit flies. *Annual Review of Entomology* **5** : 171-192.

Clark, A.R., Armstrong, K.F., Carnichel, A.E., Milne, J.R., Raghu, S., Roderick, G.K. and Yeats, D. K. (2005). Invasive phytophagous pests arising through aa recent tropical evolutionary rediation : the *Bactrocera dorsalis* complex of fruit flies. *A Review of Entomology* **50** : 293-319.

Clausen, C.P.; Clancy, D.W.; Chock, Q.C. (1965). Biological control of the oriental fruit fly (*Dacus dorsalis* Hendel) and other fruit flies in Hawaii. *United States Department of Agriculture, Technical Bulletin* No. 1322, pp.102.

Cornelius, M.L., Duan, J.J. and Messing, R.H. (2000a). Volatile host fruit odors as attractants for the oriental fruit fly (Diptera : Tephritidae). *Journal of Economic Entomology*, **93** : 93-100.

Cornelius, M.L., Nergel, L., Duan, J.J. and Messing, R.H. (2000b). Responses of female oriental fruit flies (Diptera : Tephritidae) to protein and host fruit odours in field cage and open field tests. *Environmental Entomology*, **29** : 14-19.

Cunningham, R.T. (1989a). Biology and physiology; parapheromones. In : *World Crop Pests 3(A). Fruit flies; their biology, natural enemies and control* (Ed. by Robinson, A.S.; Hooper, G.), pp. 221-230. Elsevier, Amsterdam, Netherlands.

Cunningham, R.T. (1989b). Control; insecticides; male annihilation. In : *World Crop Pests 3(B). Fruit flies; their biology, natural enemies and control* (Ed. by Robinson, A.S.; Hooper, G.), pp. 345-351. Elsevier, Amsterdam, Netherlands.

Dale, N.S. and Patel, R.K. (2010). Population dynamics of fruit flies (*Bactrocera* spp.) on guava and its correlation with weather parameters. *Current Biotica*, **4** (2) : 245-248.

David, K.J. and Ramani, S. (2011). An illustrated key to fruit flies (Diptera : Tephritidae) from peninsular India and Andaman and Nicobar Islands. *Zootexa* **302** (1) : 1-31.

De, K and Pandey, Y.D. (1988). Bionomics and some behaviour aspects of mango fruit weevil, *Sternochetus gravis* (F.). *Entomon* **13** : 17-24.

Deepa, M., Agarwal, N., Viswakarma, R., Kiran Kumari and Lal, K.M. (2009). Monitoring and weather parameters on *Bactrocera* complex through methyl eugenol traps. *Annals of Plant Protection Sciences*, **17** (2) : 332-336.

Deepa, M., Agrawal, N. and Viswakarma, R.K. (2010). Monitoring the incidence of *Bactrocera* species complex through methyl eugenol traps. *Current Advances in Agriculture Sciences,* **2** (1) : 74-79.

Dhillon, M.K., Singh, Ram, Naresh, J.S. and Sharma, H.C. (2005). The melon fruit fly, Bactrocera cucurbitae; A review of its biology and management. *Journal of Insect Science* **40** : 1-16.

Drew, R.A.I. and Hancock, D.L. (2000). Phylogeny of the tribe Dacini (Dacinae) based on morphological, distributional, and biological data. In : M. Aluja and A. L. Norrbom (Eds.), *Fruit Flies (Tephritidae) : Phylogeny and Evolution of Behavior.* CRC Press, Boca Raton, Florida, pp. 491-504.

Drew, R.A.I. (1974). The response of fruit fly species (Diptera : Tephritidae) in South Pacific area to male attractants. *Journal of the Australian Entomological Society* **13** : 267-270.

Drew, R.A.I. (1982). Fruit fly collecting. In : *Economic fruit flies of the South Pacific Region* (Ed. by Drew, R.A.I.; Hooper, G.H.S.; Bateman, M.A.) (2nd edition), pp. 129-139. Queensland Department of Primary Industries, Brisbane, Australia.

Drew, R.A.I. (1989). The tropical fruit flies (Diptera : Tephritidae) of the Atralasian and ocean regions. Memories of the Queensland museum 26 : 1-521.

Drew, R.A.I. and Hancock, D.L. (1994). The *Bactrocera dorsalis* complex of fruit flies in Asia. *Bulletin of Entomological Research : Supplement Series.* Supplement No. 2. CAB International, Wallingford, UK.

Duyck, P.F., Rousse, P., Ryckewaert, P., Fabre, F. and Quilici, S. (2004). Influence of adding borax and modifying pH on effectiveness of food attractants for melon fly (Diptera : Tephritidae). *Journal of Economic Entomology,* **97** (3) : 1137-1141.

Dwivedi, B.K., Pandey, G., Pant, H., Pandey, R.C. and Logani, R. (2002). Integrated management of guava fruit fly (*Dacus dorsalis* Hendel). *Bioved,* **13** (1/2) : 113-115.

Dwivedi, S.C., Kuldeep, Singh, S.M. and Katiyar, R.R. (2003). Seasonal incidence of insect pests associated with mango crop. *Annals of Plant Protection Sciences,* **11** : 159-160.

Economopoulos, A. P., Raptis, A., Stavopoulou-Delivoria, A. and Papadopoulos, A. (1986). Control of Dacus oleae by yellow sticky traps combined with ammonium acetate slow release dispensers. Entomologia Experimentalis et Applicata **41**(1) : 11-16.

El-Gendy, I. R. (2013). Response of the peach fruit fly, *Bactrocera zonata* (Saunders) (Diptera : Tephritidae), to synthetic food-odor lures and extent the effect of pH on attracting the fly. *Journal of Entomology,* **10** (3) : 136-146.

Elnagar, S., El-Sheikh, M., Hashem, A. and Afia, Y. (2010). Recent invasion by *Bactrocera zonata* (Saunders) as a new pest competing with *Ceratitis capitata* (Wiedemann) in attacking fruits in Egypt. *Aspects of Applied Biology,* **104** : 97-102.

Fabre, F., Ryckewaert, P., Duyck, P.F., Chiroleu, F. and Quilici, S. (2003). Comparison of the efficacy of different food attractants and their concentration for melon fly (Diptera : Tephritidae). *Journal of Economic Entomology,* **96** : 231-238.

FAO (1987). Outbreaks and new records. USA. Eradication of Oriental fruit fly. *FAO Plant Protection Bulletin* **35** :166.

Firake, D.M., Behere, G.T., Deshmukh, N.A., Firake, P.D. and Azad Thakur, N.S. (2013). Recent scenario of insect-pests of Guava in North East India and their Eco-friendly Management. *Indian Journal of Hill Farming,* **26** (1) : 55-57.

Fletcher, B.S. (1989). Ecology; movements of tephritid fruit flies. In : *World Crop Pests 3(B). Fruit flies; their biology, natural enemies and control* (Ed. by Robinson, A.S.; Hooper, G.), pp. 209-219. Elsevier, Amsterdam, Netherlands.

Fletcher, T.B. (1917). Annotated list of Indian crop pests. *Proceedings of the second Entomological Meetings,* Pusa. **2** : 307.

Foote, R.H. (1980). Fruit fly genera of sooth of the United States (Diptera : Tephritidae) Tech. Bull., No. 1600, IV : 79 pp.

Ghanim, N.M. 2013. Influence of Methyl Eugenol Diluted with Paraffin Oil on Male Annihilation Technique of Peach Fruit Fly, *Bactrocera zonata* (Saunders) (Diptera : Tephritidae) *Entomol Ornithol Herpetol,* **2** (3) : 1-6.

Galande, S.M. and Ukey, N.S. (2011). Fruit fly, *Bactrocera* species composition in Pune region of Maharashtra state. *Journal of Applied Zoological Research,* **22** (1) : 1-4.

Galande, S.M., Ukey, N.S., Jadhav, S.S., Shelke, S.S., Verma, A.K., Bhardwaj, S.P. and Gupta, P.R. (2010). Composition of fruit fly species in Pune region of Maharashtra and losses caused to guava crop in the region. *Pest Management and Economic Zoology,* **18** (1/2) : 290-294.

Gilmore, J.E. (1989). Control; sterile insect technique (SIT); overview. In : *World Crop Pests 3(B). Fruit flies; their biology, natural enemies and control* (Ed. by Robinson, A.S.; Hooper, G.), pp. 353-363. Elsevier, Amsterdam, Netherlands.

Gowda, Vijaya M. 2018. Fabrication of different indigenous fruit fly traps (methyl eugenol as attractant) and assessment of economic loss due to fruit flies in guava orchards. M.Sc. Entomology Thesis submitted to C.S.A. University of Agricultue and Technology, Kanpur.

Gupta, D. and Bhatia, R. (2001). Population fluctuations of fruit flies, *Bactrocera* spp. in sub- mountainous mango and guava orchards. *Journal of Applied Horticulture* (Lucknow), **2** (1) : 47-49.

Hampson, G.F. (1896). *In* the Fauna of British India Including Ceylon and Burma. Taylor and France, London. pp. 124-127.

Hardy, D.E. (1949). Studies in Hawaiian fruit flies. *Proceedings of the Entomological Society of Washington* **51** : 181-205.

Hancock, D.L., Osborne, R., Broughton, S. and Gleeson, P. (2000). Eradication of Bactrocera papaya (Diptera : Tephritidae) by male annihilation and protein baiting in queensland, Australia. In : Tan K H (ed.) *Area-Wide Control of Fruit Flies and Other Insect Pests. Penerbit University Sains, Malaysia, pp* 81-88

Hasseb, M. (2007). Current status of insect pest problems in guava. *Acta Horticulturae,* **735** : 453-467.

Haymer, D.S.; Tanaka, T.; Teramae, C. (1994). DNA probes can be used to discriminate between tephritid species at all stages of the life cycle. *Journal of Economic Entomology* **87** : 741-746.

Ho, K.Y., Lee, H.J., Horng, S.B. and Chen, C.C. (2003). Comparison of the effectiveness of different materials for attracting the Oriental fruit fly, *Bactrocera dorsalis* (Diptera : Tephritidae). *Plant Protection Bulletin* (Taipei), **45** (2) : 117-126.

Hocking Brain (1968). Six legged Science. Pub. Schenkman Publishing Co., Cambridge Mass.

Hong, L.J. and Hui, Y. (2005). Population dynamics of *Bactrocera dorsalis* (Diptera : Tephritidae) in Yuanjiang dry-hot valley, Yunnan with an analysis of the related factors. *Acta Entomologica Sinica,* **48** (5) : 706-711.

Hong, L. J. and Hui, Y. (2006). Effects of light, temperature and humidity on the flight activities of the Oriental fruit fly, *Bactrocera dorsalis. Chinese Bulletin of Entomology,* **43** (2) : 211-214.

Ignacimuthu, S. and Gabrial, Paul Raj M. (2009). A report on the National Symposium on non-chemical insect pest management. *Current Science,* **97** (2): 25.

J.W. and Lopez, D.E (1969). Trapping for control of the Mexican fruit fly in and

Jalaluddin, S.M., Natarajan, K. and Sadakathulla, S. (2001). Population fluctuation of the guava fruit fly, *Bactrocera correcta* (Bezzi) in relation to hosts and abiotic factors. *Journal of Experimental Zoology,* **4** (2) : 323-327.

Jiron, L.F. and Hedstrom, I. (1991). Population fluctuations of economic species of *Anastrepha* (Diptera : Tephritidae) related to mango fruiting phenology in Costa Rica. *Florida Entomologist,* **74** (1) : 98-105.

Jose, L., Cugala, D. and Santos, L. (2013). Assessment of invasive fruit fly fruit infestation and damage in Cabo Delgado Province, Northern Mozambique. *African Crop Science Journal,* **21** (1) : 21-28.

Jothi, D.B., Tandon, P.L. and Verghese, A. (1994). Hot water immersion as a guarantine treatment for Indian mangoes infested with the Oriental fruit fly, *Bactrocera* dorsalis (Mendel) (Diptera : Tephritidae). *FAO Plant Protection Bulletin* **42** (3) : 158-159.

Kalia, V. (1992). Bionomics of fruit fly *Dacus dorsalis* on some cultivars of mango and guava. *Bulletin of Entomology* (New Delhi), **33** (1/2) : 79-87.

Kalia, V. and Srivastava M.L. (1992). Ovipositional benaviour and development of oriental fruit fly Dacus (*Strumeta*) *dorsalis* Hendel on development stages of Mango fruit. *Bulletin of Entomology* **33** : 88-93.

Kapoor, V.C. (2005). Taxonomy and biology of economically important fruit flies of India. *Israel Journal of Entomology,* **35-36** : 459-475.

Kapoor, V.C. (1993). Indian fruit flies (Insecta : Diptera : Tephritidae). Oxford and IBH publishing corporation *Private Limited. New Delhi,* pp. 228.

Katsoyannos, B.I. and Papadopoulos, N.T. (2004). Evaluation of synthetic female attractants against *Ceratitis capitata* (Diptera : Tephritidae) in sticky coated spheres and Mc Phail type traps. *Journal of Economic Entomology*, **97** : 21-26.

Kawashita, T., Rajapakase, G.B.J.P. and Tsuruta, K. (2004). Population survey of *Bactrocera* fruit flies by lure trap in Sri Lanka. *Research Bulletin of the Plant Protection Service,* **40** : 83-87.

Khan, M. A. and Khaliq, M.A.A. (2003). Population of fruit fly species trapped by methyl euginol and cue lure versus infestation in guava orchards. *Pakistan Entomologist*, **25** (1) : 63-67.

Khattak, S.U., Afsar, K., Hussain, N., Khalil, S.K. and Alamzeb (1990). Annual population incidence of oriental fruit fly (*Dacus dorsalis*) Hendel in a fruit orchard at Peshawar, Pakistan. *Bangladesh Journal of Zoology*, **18** (2) : 131-138.

Knipling (1955). Possibilities of insect control of exudication through the use of sexually sterile males. *Journal of Economic Entomology* **48** (4) : 459-462.

Kumar, C.H.R. and Viraktamath, S. (2007). Efficacy of protein food baits in attracting female fruit flies in guava and mango orchards. *Pest Management and Economic Zoology*, **15** (1) : 13-18.

Kumar, Pankaj (2018). Efficacy of some attractants for monitoring of fruit flies, *Bactrocera* spp. prevalent in Guava and Mango through trapping technique. Ph.D. Thesis submitted to CSA University of Agriculture and Technology, Kanpur.

Kumar, S., Patel, C.B. and Bhatt, R.I. (1997). Studies on seasonal cyclicity of *Bactrocera correctus* Bezzi in mango and sapota orchards using methyl eugenol trap. *Gujarat Agricultural University Research Journal*, **22** : 68-74.

Kumar, V. and Agarwal, M.L. (1998). Efficacies of different bait combinations against oriental fruit fly, *Bactrocera dorsalis* (Hendel). *Journal of Research,* Birsa Agricultural University, **10** : 83-86.

Latif, A. and Abdullah, K. (2005). Population dynamics and management of *Bactrocera* sp. (Diptera : Tephritidae) infesting mango. *Pakistan Entomologist*, **27** (2) : 69-72.

Leblenk, L. and Putoa, R. (2000). Fruit flies in French Polynesia and Pitcairn Islands. *Plant Protection Services Secretariat of the Pacific Community*. 29 pp. 4.

Lee, Fngming and Lu, F.M. (1997). Ovipositional preference of the melon fly, *B. cucurbitae* Coq. (Diptera; Tephritidae). Tests of host plant and colour. *Chinese Journal of Entomology,* **17** (4) : 237-343.

Lee, W.Y.; Cheng, C.C. and Lin, M.J. (1990). Morphological studies of the rectal papillae of the male oriental fruit fly, *Dacus dorsalis* Hendal, using scanning electron microscopy. *Bulletin Institutional Zoology,* Academia sinica, **29** (3) : 195-205.

Liqunido, N.T.; Cunningham, R.T.; Nakagawa, S. and Uchida, G. (1990). Survey of *Dacus cucurbitae* coq. (Diptera : Tephritidae) infestations in the cultivated and weedy forms of *Momordica charantia* L. (Cucurbitae), *Proceeding of the Hawaiian Entomological Society* No. 30 : 31-36.

Madhura, H.S. (2001). Management of fruit flies (Diptera : Tephritidae) using physical and chemical attractants. *M.Sc. (Agri.) Thesis*, University of Agricultural Sciences, Bangalore, p. 80.

Madhura, H.S. and Viraktamath, C.A. (2003). Efficacy of different traps in attracting fruit flies (Diptera : Tephritidae). *Pest Management in Horticultural Ecosystems*, **9** : 153-154.

Madhura, H.S. and Virakthamath, C.A. (2001). Selective respond of fruit flies (Diptera : Tephritidae) to traps at different heights. In A. Verghese and P.R. Reddy (Eds.) Proceedings of 2nd International Symposium on Integrated Pest Management (IPM) in Horticultural Crops : Different New Molecules, Biopesticides and Horticulture Ecosystem, Bangalore pp 29-30.

Magid, F.M.A., Ali, E.E., Khair, A.M. and Mahmoud, M.E.E. (2012). Life cycle study of *Bactrocera invadens*, *Ceratitis cosyra* and *Ceratitis capitata* (Tephritidae : Diptera) and assessment of their damage to mango, guava and citrus in Shendi Area, Sudan. *Persian Gulf Crop Protection*, **1** (2) : 22-27.

Mahmoud, M.E.E., Kambal, M.A.O. and Abukashawa, S.M.A. (2012). Prospects of using protein hydrolastes for trapping and monitoring major fruit flies (Tephritidae : Diptera) in Sudan. *Persian Gulf Crop Protection,* **1** (2) : 6-14.

Makhmoor, H.D. and Singh, S.T. (1998). Effective concentration of methyl eugenol for the control of guava fruit fly, *Dacus dorsalis* Hendel in guava orchard. *Annals of Plant Protection Sciences,* **6** (2) : 165-169.

Mann G.S. (1996). Seasonal incidence and build up of Betrocera dorsalis Hendel on Mango in Punjab. *Journal of Insect Science* I : 129-132.

Mann, G.S. (1991). Seasonal incidence and build-up of *Bactrocera dorsalis* Hendel on mango in Punjab. *Journal of Insect Science,* **9** (2) : 129-132.

Manzar, Abu and Srivastava, J.P. (2004). Population fluctuation of fruit flies, Bactrocera spp infesting bittar dourd in Central Uttar Pradesh. *Progressive Horticulture* **36** (1) : 146-149.

Medeiros, J.G.F., Malta, A.O., Costa, N.P., Araujo, R.C. and Araujo, E.L. (2011). Substance attractive the monitory of fruit flies the guava and mango in municipality Bananeiras-PB [Portuguese]. *Revista Verde de Agroecologia e Desenvolvimento Sustentavel,* **6** (5) : Artigos 213-219.

Munj, A.Y., Salvi, B.R., Jalgaonkar, V.N., Damodhar, V.P. and Narangalkar, A.L. (2013). Rakshak methyl eugenol traps : an ideal tools developed by Dr. B.S. Konkan Krishi Vidyapeeth, Dapou to minimize the mango fruit fly incidence (*Bactrocera dorsalis* Hendel) during unseasonal rains. *Annals of Plant Physiology,* **27** (1/2) : 54-57.

***Nagappan, K., Kamalnathan, S., Santharaman, T. and Ayyapan, M. (1971).** Insecticidal trials for the control of the melon fly, *Dacus cucurbitae* Coq. infesting snake gourd, *Trichosanthes anguina. J. Madras Agril.,* **58** : 684–690.

Nagaraj, K.S. Jaganath, S. Srikanth, L.G. and Mahamadtoufeeq Husainnaik (2014a). Attraction of fruit fly to different quantities of methyl eugenol in mango orchard. *Trends in Biosciences,* **7** (11) : 1113-1115.

Nagaraj, K.S. Jaganath, S. Swamy, G.S.K. (2014b). Effect of protein food baits in attracting fruit flies in mango orchard. *Asian Journal of Horticulture,* **9** (1) : 190-192.

Nagaraj, K.S., Jaganath, S., Bawoor, S., Srikanth, L.G. and Mahamadtoufeeq, H. (2014c). Incidence of fruit fly (*Bactrocera* spp.) in different mango orchards and varieties. *Trends in Biosciences,* **7** (11) : 1072-1074.

Nan, W., Mi KhinMi, Oo ThiTar, Win KyawKyaw, Park JinYoung and Park JongKyun (2014). Occurrence of fruit flies (Diptera : Tephritidae) in fruit orchards from Myanmar. *Korean Journal of Applied Entomology,* **53** (4) : 323-329.

Nandre, A.S. and Shukla, A. (2013). Varietal evaluation of sapota for resistance to Oriental fruit fly, *Bactrocera dorsalis* (Hendel). *Pest Management in Horticultural Ecosystems,* **19** (2) : 164-168.

Narayanan, E.S. and Batra, H.N. (1960). Fruit flies and their control. Pub. ICAR, New Delhi, pp. 10-65.

Nboyine, J. A., Billah, M. K. and Afreh-Nuamah, K. (2012). Species range of fruit flies associated with mango from three agro-ecological zones in Ghana. *Journal of Applied Biosciences,* **52** : 3696-3703.

Panday, A.K., Kumar, A., Nath, P. and Rai, A.B. (2012). Ecofriendly management of fruit fly infesting bitter gourd. Journal of Plant Protection and Environment, **9** (2) : 28-31.

Pena, J.E, Mohyuddin and Wysoki, M. (1998). A Review of the pest management situation in Mango agro ecosystems. *Phytoparasitica* **26** (2) : 129-148.

Pruthi, H. S. and Bhatia, H.L. (1938). A preliminary annotated list of fruit pests of the north- west Frontier province. *Miscellaneous Bulletin Indian council of Agricultural Research,* New Delhi, No. **19** : 1-22.

Qureshi, Z.A., Hussain, T and Siddiqui, Q.H. (1991). Relative preference of mango varieties by *Dacus zonatus* (Saunders) and *D. dorsalis* Hendel. *Pakistan Journal of Zoology* **23**(1) : 85-87.

Rahmann, G. and Aksoy, U. (2014). Proceeding of the 4th ISOFAR Scientific conference building organic bridges, at the organic World congress 2014, 13-15 oct., Istanbul, Turkey (e print Id 23314).

Rai, S., Shankar, U., Bhagat, R.M. and Gupta, S.P. (2008). Population dynamics and succession of fruit fly on sub-tropical fruits under rain fed condition in Jammu Region. *Indian Journal of Entomology,* **70** (1) : 12-15.

Rajitha A.R. and Viraktamath S. (2005a). Monitoring fruit flies (Diptera : Tephritidae) in guava orchard in Dharwad. *Kamataka Journal of Agriculture Science* **19** : 35-39.

Rajitha, A.R. and Viraktamath, S. (2006a). Monitoring of fruit flies (Diptera : Tephritidae) in guava orchard at Dharwad, Karnataka. *Karnataka Journal of Agricultural Sciences,* **19** (1) : 45-49.

Rajitha, A.R. and Viraktamath, S. (2005b). Evaluation of protein food baits in attracting female fruits flies in guava and mango orchards. *Pest Management in Horticultural Ecosystems,* **11** : 1-5.

Rajitha, A.R. and Viraktamath, S. (2006b). Investigations on the population dynamics of fruits flies in mango orchard at Dharwad. *Karnataka Journal of Agricultural Sciences,* **19** : 134-137.

Ramani, S. (1997). Biosystematic studies on fruit flies (Diptera : Tephritidae) with special reference to the fauna of Karnataka and Andaman and Nicobar. Ph.D. Thesis, *University of Agriculture Science, Bangalore pp* 214.

Robacker, D.C. (1992). Effects of shape and size of coloured traps on attractiveness to irradiated, laboratory strain Mexican fruit flies (Diptera : Tephritidae), *Florida Entomologist* **75**(2) pp. 230-241.

Roomi, M.W., Abbas, T., Shah, A.H., Robina, S., Qureshi, S.A., Hussain, S.S. and Nasir, K.A. (1993). Control of fruit flies (*Dacus* spp.)

Ravikumar, Ch. (2005). Studies on fruit flies in guava and mango orchards with special reference to their management through mass trapping. *M.Sc. (Agri.) Thesis,* University of Agricultural Sciences, Dharwad, p.90.

Ravikumar, P. and Viraktamath, S. (2007). Attraction of female fruit flies to different protein food baits in guava and mango orchards. *Karnataka Journal of Agricultural Sciences,* **20** (4) : 745-748.

Reghunath, P. and Indra, M. (2000). Fruit fly trap made from Basil. ILEIA News letter **9** : 31.

Rejirani, O.P., Paul, A. and Jiji, T. (2012). Field evaluation of methyl eugenol trap for the management of Oriental fruit fly, *Bactrocera dorsalis* Hendel (Diptera : Tephritidae) infesting mango. *Pest Management in Horticultural Ecosystems,* **18** (1) : 19-23.

Saputra D.W and Marmaini (2016). Trap colour effect of fruit flies in cropping lime village sigma maura enim district gelumbang, *International Research Journal of natural Sciences,* **4**, pp.1-14.

Sarada, G., Maheswari, T.U. and Purushotham, K. (2001b). Seasonal incidence and population fluctuation of fruit flies in mango and guava. *Indian Journal of Entomology,* **63** : 272-276.

Sarode, S.V., Lalitha, P. and Krishnamurthi, P. (1981). Residues of fenthion in/on musk melon. *Journal of Entomological Sciences,* **5** : 179-181.

Satarkar, V.R., Krishnamurthy, S.V., Falerio, J.R., Verghese, A. and Stonehouse, J.M. (2009). An assessment of methyl eugenol dispensers and fruit fly species complex in orchard agro-ecosystems of Goa, India. *Pest Management in Horticulture Ecosystems,* **12** (2) : 161-163.

Seewooruthun, S.I., Permalloo, S. and Sookar, P. (2005). Evaluation of food-based attractants for fruit fly trapping in Mauritius. In **:** *Proceedings of a final Research Coordination Meeting organized by the Joint FAO/IAEA Programme of Nuclear Techniques in Food and Agriculture,* 5-7 May 2005, Vienna, pp. 202-237.

Seewooruthun, S.I., Sookar, P., Permalloo, S., Joomaye, A., Alleck, M., Gungah, B. and Soonnoo, A.R. (2001). An attempt at the eradication of the oriental fruit fly, *Bactrocera dorsalis* (Hendel) from Mauritius. In : *Annual Report of the Agricultural Services of the Ministry of Agriculture, Fisheries and Co-operatives,* pp.1-10.

Sewoosunkur Gopaul, Zenz, N., Price, N., Price, N. S. and Seewooruthun, I. (2000). Local production of protein bait for fruit fly monitoring and control. *In* : Proceedings of the Indian Ocean Commission, Regional Fruit Fly Symposium, Flic en Flac, Mauritius, 5th-9th June, pp. 41-47.

Sharma, I.D., Kumar, S., Chandel, R.S. and Patyal, S.K. (2011). Evaluation of Drek, *Melia azadirach* for the management of fruit flies, *Bactrocera tau* in tomato. *Journal of Bio pesticides,* **4** (1) : 1- 5.

Shukla R.P. and Prasad V.G. (1985). Population fhictuation of the oriental fruit fly Dacus dorsalis Hendel in relation to hosts and abiotic factors. *Tropical pest Management* **31** : 273-275.

Shukla R.P., Tandon, P.L. and Suman, G.L. (1984). *Baculovirus* a new pathogen of mango nut weevil, *Sternochetus mangiferae* (Fabr.) (Coleoptera : curculionidae). *Current Science,* **53** (11) : 593-594.

Singh, A and Lata, P. (1989). The mole terminalia of seven species of genus *Dacus. Bulletin of Entomology* **30** (1) : 56-64.

Singh, G. (1991). Loss assessment, ecology and management of Mango fruit fly *Dacus* sp. *Acta Horticulture* **291** : 425-436.

Singh, M., Gupta, D. and Gupta, P.R. (2013). Population suppression of fruit flies (*Bactrocera* spp.) in mango (*Mangifera indica*) orchards. *Indian Journal of Agricultural Sciences,* **83** (10) : 1064-1068.

Singh, R. (2008). Studies on varietal susceptibility of oriental fruit fly, *Bactrocera dorsalis* (Hendel) on guava and its attraction to different poison baits. *Asian Journal of Bio Science,* **3** (2) : 330-332.

Singh, S. (1980). Studies on the *Apsylla cistelata* (Buckton) and its control. M.Sc. (Ag) thesis G.B. Pant University of Agriculture & Technology, Pantnagar, Uttarakhand.

Singh, S. and Sharma, D.R. (2013). Management of fruit flies in rainy season guava through male annihilation technique using methyl eugenol based traps. *Indian Journal of Horticulture,* **70** (4) : 512-518.

Sivinski, J. (1990). Coloured spherical, traps for capture of Caribbean fruit fly, *Anastripha suspensa. Florida Entomologist,* **62** : 123-128

Solangi, B., Sultana, R., Ahmed, N., Waganand, M.S. and Abbassi, A.R. (2014). Seasonal monitoring of fruit fly, *Bactrocera zonata* (Saunders) and its parasitoids *Trybliographa daci* on guava fruit from Sindh Pakistan. *Journal of Biodiversity and Environmental Sciences (JBES),* **4** (5) : 222-232.

Sood, P. and Nath, A. (1998). Evaluation of insecticide baits for the control of fruit fly *Bactrocera tau* (Walker) in the mid hills of Himachal Pradesh. *Journal of Hill Research,* **11** : 171-177.

Srivastava, R.P. and Tandon, P.L. (1980). New records of parasites and predators of important insect pests of mango. *Entomon.* **5** (3) : 243-244.

Srivastava, R.P. and Tandon, P.L. and Verghese, A. (1982). Evaluation of insecticides for the control of mango shoot gall psyla, *Apsylla cistelata* (Buckton) (Psyllidae, Homoptera) *Entomon.* **1** : 281-284.

Stibick, J.N.L. (2004). Natural enemies of true fruit flies (Tephritidae). USDA Marketing and Regulatory Programs, animal and plant health, Inspection, Service, Plant Protection and Quarentine. pp. 3-30.

Stonehouse, J. (2001). An over view of fruit fly research know ledge and needs in the Indian Ocean region. In : proceedings of the second national symposium on Integrated Pest Management (IPM) in Horticultural crops. *New molecules, Biopesticides and Environment,* Bangalore pp. 21-23.

Stonehouse, J., Muhammad, A., Zia, Q., Mumfores, J., Poswal, Anand Mahmood, R. (2002). Single-killing-point field assessment of bait and lure control of fruit flies (Diptera : Tephritidae) in Pakistan *Crop Protection,* **21** (8) : 651-659.

Stonehouse, J.M., Verghese, A., Mumford, J.D., Thomas, J., Jiji, T., Faleiro, R., Patel, Z.P., Jhala, R.C. Patel, R.K., Shukla, R.P., Satpathy, S., Singh, H.S., Amerika Singh and Sardana, H.R. (2005). Research conclusions and recommendations for the on-farm IPM of tephritid fruit flies in India. Pest Management in Horticultural Ecosystems, **11** (2) : 172-180.

Stonehouse, J.M., Mumford, J.D., Verghese, A., Shukla, R.P., Satpathey, S., Singht, H.S., Jiji, T., Thomas, J., Patel, Z.P., Jhala, R.C., Patel, R.K., Manzar, A., Shivalingaswamy, T.M., Mohantha, A.K., Nair, B., Vidya, C.V., Jagadale, V.S., Sisodiya, D.B. and Joshi, B.K. (2007). Village level area wide fruit fly suppression in India : Bait application and male annihilation at village level and farm level. *Crop Protection,* **26** (5) : 788-793.

Sunandita and Gupta, D. (2001). Testing of boric acid and protein hydrolysate bait mixture against fruit fly, *Bactrocera tau* Walker. *Indian Journal of Entomology,* **63** : 125-129.

Sureshbabu, K. and Viraktamath, S. (2003). Species diversity and population dynamics of fruit flies (Diptera : Tephritidae) on mango in Northern Karnataka. *Pest Management and Economic Zoology,* **11** : 103-110.

Tandon, P.L. (1995). Integrated insect pest management in mango. *In* Proceedings National Symposium of Integrated Pest Management and Environment Organised by Plant Protection Association of India and FAO at Madras, Feb. 2-4, 1995.

Tandon, P.L. and Lal, B. (1978). The mango coccid, *Rastrococcus iceryoides* (Green) (Homoptera : Coccidae) and its natural enemies *Current Science,* **47** : 467-468.

Tandon, P.L. and Verghese, A. (1985). World list of insect, mite and other pests of mango. Technical Document No. 5, IIHR, Bangalore pp 22.

Thakur, M., Gupta, D. and Singh, M. (2013). Monitoring of fruit fly populations using parapheromone in peach orchards. *Journal of Insect Science* (Ludhiana), **26** (1) : 34-41.

Tharanathan, R.N., Yashoda, H.M. and Prabha, T.N. (2006). "Mango (*Mangifera indica* L.), 'The King of Fruits'- An Overview". *Food Reviews International,* **22** (2) : 95-123.

Thomas, D.B., Holler, T.C., Heath, R.R., Salinas, E.J. and Moses, A.L. (2001). Trap-lure combinations for surveillance of *Anastrepha* fruit flies (Diptera : Tephritidae). *Florida Entomologist,* **84** : 344-351.

Toth, M., Voigt, E., Baric, B., Pajac, I., Subic, M., Baufeld, P. and Lerche, S. (2014). Importance of application of synthetic food lures in trapping of *Rhagoletis* spp. and *Strauzia longipennis* Wiedemann. *Acta Phytopathologica et Entomologica Hungarica,* **49** (1) : 25-35.

Tsurata, K. and White, I.M. (2001). Eleven new species of genus *Bactrocera Macquart* (Diptera : Tephritidae) from Sri Lanka. *Entomological Science* 4 : 69-87.

Toorani, A.H and Abbasipour, H. (2017). Effect of geographical directions height and the colour of yellow traps in capture of Mediterranean fruit fly,, *Ceratitis capitate* (Wiedemann, 1824) incitrus orchards*, Acta agriculturae Slovenica*, 109-3, pp 561-575.

Ukey, N.S., Chandele, A.G., Wagh, S.S. and Bansode, G.M. 2013. Species composition of fruit flies, *Bactrocera* spp. (Diptera : Tephritidae) infesting guava in Maharashrtra. *Pest Management in Horticultural Ecosystems,* **19** (2) : 242-244.

Ullah, F., Badshah, H., Alam, K., Khan, A., Zeb, Q. and Mashwani, M.A. (2012). Evaluation of various insecticides as lure toxicants for fruit fly management in guava orchards. *Sarhad Journal of Agriculture,* **28** (4) : 617-620.

Vargas R.I. Stark J.D. and Nishida T. (1989). Abundance distribution and dispersion indices of the oriental fruit fly and melon fly (Diptera : Tephritidae) in Kavai Hawallon Islands. *Journal of economic Entomology* 82 : 1609-1516.

Vayssieres, J.F., Sinzogan, A., Adandonon, A., Rey, J.Y., Dieng, E.O., Camara, K., Sangare, M., Ouedraogo, S., Hala, N., Sidibe, A., Keita, Y., Gogovor, G., Korie, S., Coulibaly, O., Kikissagbe, C., Tossou, A., Billah, M., Biney, K., Nobime, O., Diatta, P., N'Depo, R., Noussourou, M., Traore, L., Saizonou, S. and Tamo, M. (2014). Annual population dynamics of mango fruit flies (Diptera : Tephritidae) in West Africa : Socio-economic aspects, host phenology and implications for management. *Fruits (Paris),* **69** (3) : 207-222.

Vayssieres, J.F., Vanniare, H., Gueye, P.S., Barry, O., Hanne, A.M., Korie, S., Niassy, A., Ndiaye, M. and Delhove, G. (2011). Preliminary inventory of fruit fly species (Diptera, Tephritidae) in mango orchards in the Niayes region, Senegal, in 2004. *EDP Sciences S.A.,* **66** (2) : 91-107.

Vergas, R.I., Stark, J.D. and Nishida, T. (1990). Population dynamics, habit preference and seasonal distribution pattern of oriental fruit fly and melon fly (Diptera and Tephritidae) in agriculture. *Environment Entomology* **19** (6) : 1820-1828.

Verghese, A. and Sudhadevi, K. (1998). Relation between trap catch of *Bactrocera dorsalis* Hendel and abiotic factors. *In : Proceedings of First National Symposium on Pest Management in Horticultural Crops,* Bangalore. pp. 15-18.

Verghese, A., Madhura, H.S., Jayanthi, P.D.K. and Stonehouse, J.M. (2002). Fruit flies of economic significance in India, with special reference to *Bactrocera dorsalis* (Hendel). *In* : *Proceedings of the 6th International Symposium on fruit flies of economic importance,* Stellenbosch, South Africa, 6-10 May 2002, pp. 317-324.

Verghese, A., Nagaraju, D.K., Madhura, H.S., Jayanthi, P.D.K. and Sreedevi, K. (2006). Wind speed as an independent variable to forecast the trap catch of the fruit fly (*Bactrocera dorsalis*). *Indian Journal of Agricultural Sciences*, **76** (3) : 172-175.

Verma, Shyam Kumar (2013). Bio-efficacy of indigenous plant extracts on population suppression of guava fruit fly, *Bactrocera correcta* (Bezzi) infesting guava. Ph.D. Thesis submitted to CSA University of Agriculture and Technology, Kanpur. 93pp.

Viraktamath, S. and Sureshbabu, K. (2004). Species composition and population dynamics of fruit flies (Diptera : Tephritidae) on Guava. *South Indian Horticulture*, **52** (1-6) : 317-323.

Wakabayashi, N. and Cunningham, R. T. (1991). Four-component synthetic food bait for attracting both sexes of the melon fly (Diptera : Tephritidae). *Journal of Economic Entomology*, **84** (6) : 1672-1676.

White I.M. and Elson-Harris, M.M. (1992). Fruit flies of economic significance : Their identification and Bionomics. CAB International. Wallingbord U.K. 601pp.

Zhang- Zhiying; Yu, H.D., Yuping, She, Zhang- zy and Dy, He. (1995). The population dynamics of Oriental fruit fly in Yunnan province. *Acta Phytophylacia Sinica* **22** (3) : 211-216.

❑❑❑